T0337028

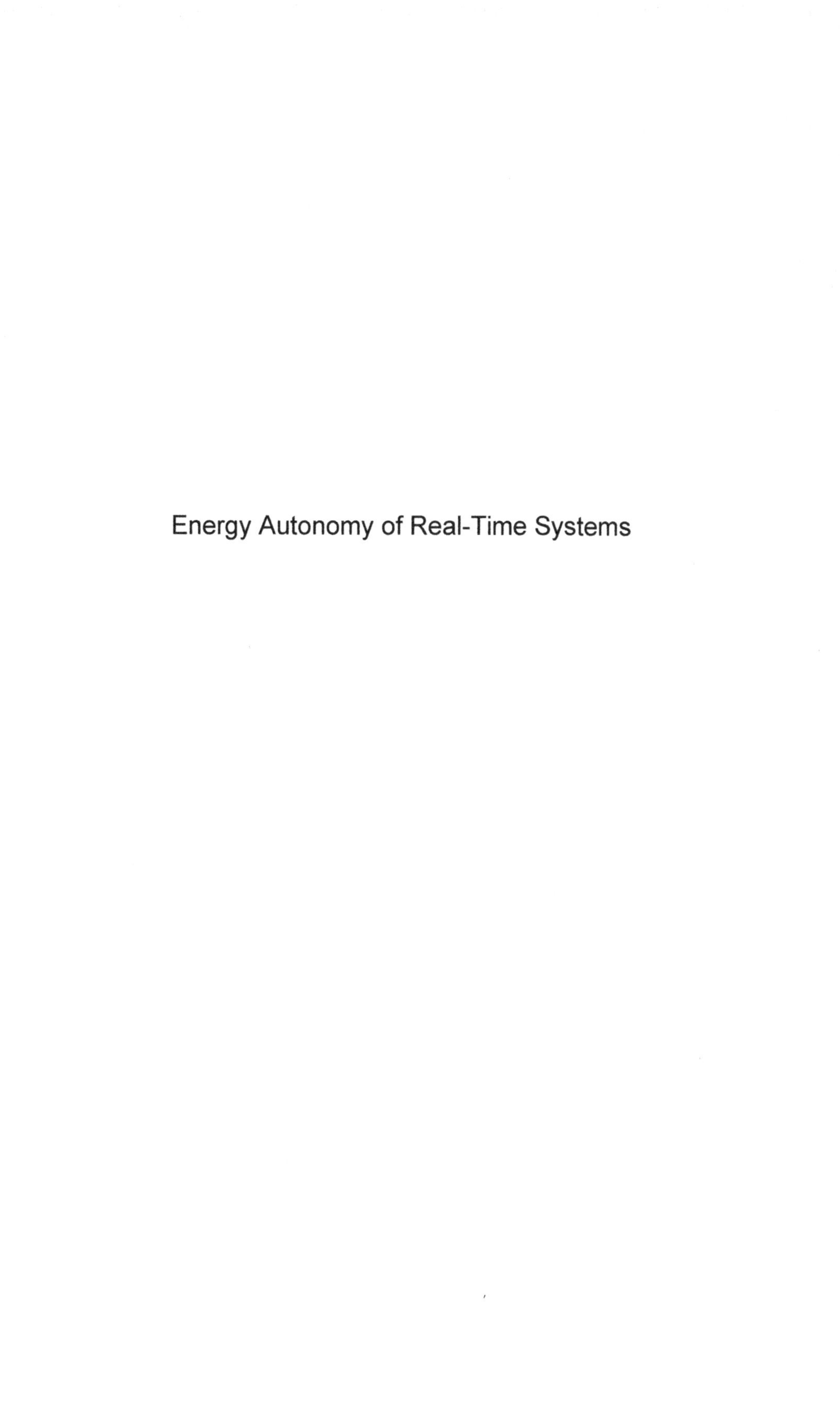

Energy Autonomy of Real-Time Systems

Energy Management in Embedded Systems Set

coordinated by
Maryline Chetto

Energy Autonomy of Real-Time Systems

Maryline Chetto
Audrey Queudet

First published 2016 in Great Britain and the United States by ISTE Press Ltd and Elsevier Ltd

ISTE Press Ltd
27-37 St George's Road
London SW19 4EU
UK

www.iste.co.uk

Elsevier Ltd
The Boulevard, Langford Lane
Kidlington, Oxford, OX5 1GB
UK

www.elsevier.com

British Library Cataloguing-in-Publication Data
A CIP record for this book is available from the British Library
Library of Congress Cataloging in Publication Data
A catalog record for this book is available from the Library of Congress
ISBN 978-1-78548-125-3

Printed and bound in the UK and US

Contents

Introduction

Digital communication equipment is becoming increasingly important in our daily lives. Managing electrical energy is a crucial problem for these devices. For cell phones equipped with rechargeable batteries, the goal is to minimize energy consumption in order to maximize the battery life, i.e. the period between two charges. However, for a significant number of devices and, in particular, last-generation embedded systems such as wireless sensor nodes, human intervention is restricted or even impossible. This is especially relevant for sensors that are inaccessible or located in hostile zones, or for networks with large numbers of nodes. These electronic systems operate using a small reserve of energy, in the form of a battery and/or a super-capacitor that continuously self-charges from a renewable source of energy.

This technology, known as "energy harvesting", consists of the process of generating electricity by conversion from another form of energy using well-understood physical principles such as piezoelectricity, thermoelectricity, etc. We can use this technology to design autonomous wireless systems with lifetimes spanning several years or even decades, limited only by the longevity of material components. It will be indispensable for the development of applications in embedded electronics, the civil sector (medicine, environmental protection, etc.) and the military sector (surveillance of enemy zones, embedded devices for soldiers, etc.).

Software programs integrated in these embedded systems are subject to real-time execution constraints. They must process and transmit information, in particular physical data derived from sensors, over wireless connections within a fixed time. Scheduling different activities on the processor subject to these time constraints is the main challenge of all real-time computer systems. Over the last four decades, research has mainly focused on devising scheduling solutions for physical architectures without energy constraints. There are therefore a number of scientific and technological hurdles that must be overcome before real-time systems can be fully autonomous from the perspective of energy management. In particular, we must revisit the question of scheduling while bearing in mind the additional constraint of a limited and variable energy supply.

This book examines the question of real-time scheduling in embedded systems that rely on renewable energy harvesting. It could serve as the basis for a lecture course on this topic within the context of a Bachelor's or Master's program. It could also be used as a reference for engineers and scientists working towards designing and developing software for connected devices. Many of these devices operate in real time. They require specific scheduling solutions in order to meet both energy and time constraints.

1

Real-time Computing

This chapter gives an introduction to the topic of real-time computer systems. It aims to present the basic principles and concepts associated with the design and utilization of these systems. We will provide a description of the various representative classes of real-time applications according to their time constraints, and we will examine the consequences of violating these constraints both on the system itself and on the environment that it controls. We will also discuss the way that these real-time systems can be scheduled to adapt to the process that they control while meeting the necessary time constraints. Using a few illustrative examples from avionics, multimedia and medicine, we will gain an explicit view of the link between the functional specifications of these applications and their temporal constraints. We will then move on to a discussion of the specific details of real-time operating systems from the perspective of task management, memory management and interrupt handling. The chapter ends with the presentation of a few examples of real-time operating systems, with a special focus on those that are particularly relevant in the context of embedded systems.

This chapter is organized as follows:

- What are real-time systems?
- Classification of real-time systems
- Typical examples of real-time systems

– Real-time operating systems: what are their special features?

– Examples of real-time operating systems for embedded systems

1.1. What are real-time systems?

1.1.1. *The concept of "real-time"*

The property of *real-time* describes the capacity of a computer system to react online, within certain time constraints, to the occurrence of asynchronous events arising in the external environment. The concept of *real-time* is directly linked to the ability of the system to achieve a response time (or reaction time) that may be considered adequate in the context of the controlled environment. The response time must be sufficiently small compared with the rate of change of the prompts emitted by the external environment.

The definition of *real time* widely adopted by the scientific and industrial communities is due to Stankovic [STA 88]: *the correctness of the system depends not only on the logical result of the computation but also on the time at which the results are produced*. The system must therefore be capable of reacting sufficiently quickly for the reaction to be meaningful. Consequently, real-time applications generally involve activities associated with time constraints. Strict processing deadlines are one of the most common examples.

1.1.2. *Features and properties of real-time systems*

A real-time system may be defined as a system that both interacts with an external environment that evolves over time and is responsible for some type of functionality in connection with this environment (see Figure 1.1), often with limited resources. These resources are usually material components such as physical memory, the processor, the communication channel, etc. We will see that in the case of autonomous real-time systems, energy is a critical resource that is only available in a quantity that is not only limited, but is also subject to fluctuation.

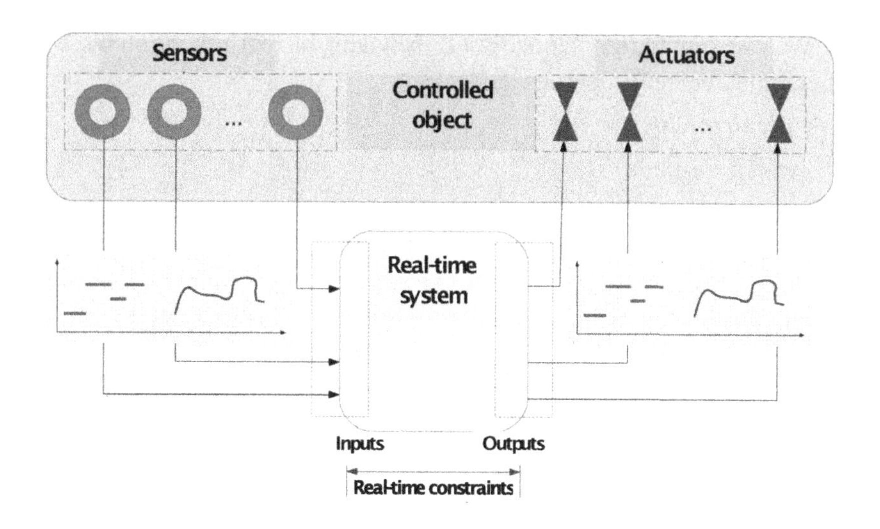

Figure 1.1. *A real-time system within its environment*

Real-time systems are defined by their integrated character: they are a component of a larger system that interacts with the physical world. This is often the primary source of the complexity of real-time systems. The physical world generally operates *non-deterministically*: events unfold in an unsynchronized manner, concurrently, in an unpredictable order.

In this context, the following two constraints must nonetheless be met in order for the real-time system to be functional:

– *logical correctness*;

– *timeliness.*

The system must not only produce the correct output as a function of its input, but must do so at the right moment in time. The *response time* (or *reaction time*) of a real-time system corresponds to the period between system input and system output during which the system processes the input. The specified response time therefore depends on the timescale of change in the environment.

We can summarize the necessary behavior of real-time systems by three fundamental properties [BON 99]:

– *predictability*;

– *determinism*;

– *reliability*.

Indeed, the system activity must be planned and executed within the specified time constraints. In order to guarantee this, the system must be designed to meet the *worst-case scenario*. The role of determinism is to remove all uncertainty in the behavior of individual activities, including the interaction of these activities. Non-determinism can arise from computational load, input/output, interrupts, etc. For the property of reliability to be met, the material and software components must be reliable when operated in real time. Accordingly, real-time systems are generally designed to be fault-tolerant.

1.2. Classification of real-time systems

1.2.1. *Hard, soft or firm: which time constraints?*

The traditional classification of real-time systems is generally based on three *levels of real-time constraints*, which are described below.

Real-time systems with strict constraints or *hard real-time systems* [SER 72, LIU 73, BLA 76] are systems that must imperatively fulfill their time constraints in order to function normally. The logical and temporal performance of these systems must be predictable [LAW 83]. A failure of the system to meet a single time constraint is considered an unacceptable fault, resulting in catastrophic damage to the controlled environment. A single time fault, or in other words a single missed constraint, could cause unacceptable loss of human life, material damage or economic loss. These computer systems, which perform processing in strict or hard real time, are most commonly encountered in aeronautics, aerospace, robotics and nuclear or chemical plant monitoring, etc. They are sometimes also described as critical real-time systems.

Real-time systems with relative constraints or *soft real-time systems* [HOR 74, LAG 76, GRA 79] can tolerate temporary violations of time constraints. The system performance is degraded without incurring significant damage to the controlled environment, and without compromising the system integrity. For example, a program that finishes execution after the deadline might result in a less precise calculation or a reduced data refresh rate, etc.

At the boundary between the previous two levels of constraints, we find *real-time systems with firm constraints* or *firm real-time systems* [BER 01]. Processing tasks described as firm real time are allowed to violate time constraints in certain pre-specified conditions. The observance of time constraints can be measured in the form of a probabilistic parameter called the *Quality of Service* (QoS). This parameter is directly linked to a service offered by the system and/or the overall behavior of the system. Tolerated missed deadlines result in a deterioration of the QoS [ABE 98]. The system performance is locally or globally degraded while maintaining a *safe* state. In the context of *firm* constraints, defining the system needs in terms of the QoS is necessary to ensure that the service provided meets the requirements specified by the end-user. Applications with *firm* real-time constraints are mostly encountered in the domains of multimedia, automated control and remote surveillance.

Consider for example an anti-lock braking system in a car [HAM 97]. In this system, one cyclically executed real-time task is to determine the moment at which braking begins by periodically inspecting the sampled speed of rotation of each wheel. The goal of the scheduler in this system is to limit the number of consecutive jobs arising from this same periodic task that miss their deadlines. A similar example is given by a multimedia application. The process of transmitting pixel packets is allowed to degrade on a saturated network by omitting certain jobs. This results in a deteriorated image after reconstruction. In this context, it is clear that the acceptable threshold for the number of packets transmitted must be monitored and specified, for example over a given period.

1.2.2. *Event-triggered versus time-triggered: which scheduling?*

There are two different activation paradigms for real-time systems in the context of applied tasks [KOP 91]: the *time-triggered* model and the *event-triggered* model. The *time-triggered* model is a purely periodic model of interaction: all applied tasks are activated at regular intervals in time. In the *event-triggered* model, asynchronous events associated with the occurrence of input data submitted to the real-time system instead trigger the activation of tasks.

The *time-triggered* paradigm is well suited for systems that are fully characterized before deployment, i.e. systems with a finite number of possible observable states for the environment. The task parameters must also all be known beforehand (in particular their arrival times), as well as the global system parameters, over the entire lifetime of the system. The scheduling sequence of tasks, which is established offline, then simply relies on a dispatcher that executes each task according to a set of rules laid down statically in a scheduling table. This deterministic approach has many advantages:

– low dispatcher execution overhead;

– good utilization of processor resources;

– easier integration of non-temporal constraints (energy, cost).

However, anything that is not *completely known beforehand* cannot be managed (zero flexibility), and this approach is only perfectly suitable in periodic environments.

As soon as some level of flexibility is required in the system, the *event-triggered* paradigm is preferred. Real-time tasks are no longer required to be fully specified. Scheduling decisions are established online according to a certain policy. Online scheduling allows higher flexibility to be attained. However, one must bear in mind that anything that is not fully known beforehand cannot be guaranteed at the time of execution, regardless of the scheduling policy. Only events that have been associated with a certain task can be managed. The predictability of the model is therefore limited. Furthermore, even if the model does not assume a periodic environment, this hypothesis will need to be reintroduced to establish any offline guarantees.

1.3. Typical examples of real-time systems

1.3.1. *Avionic systems*

Almost all fighter aircraft are equipped with ejector seats. This security system allows the pilots and the crew to escape a situation in which the aircraft cannot be controlled or is about to crash. The control system of this kind of device clearly belongs to the category of systems with strict real-time constraints. Missing a time constraint could have drastic consequences for the pilot.

Figure 1.2. *ACES II ejector seat in an F-16 Falcon*

Consider, for example, the ejector seat ACES II (Advanced Concept Ejection Seat II) installed in the F-15 Eagle, F-16 Falcon, F-22 Raptor, A-10 Thunderbolt II, B-1B Lancer, B-2 Spirit and F-117 Nighthawk models (see Figure 1.2). The seat is equipped with a barometric system that measures the flight conditions (altitude, speed) at the moment of ejection, allowing one of three operating modes to be selected, which are:

– Mode 1: low speed (<250 knots or <288 mph) and low altitude (<15,000 feet). In this mode, the main parachute is deployed as soon as the seat detaches from its mount. The pilot parachute (a small parachute less than 1 m diameter designed to stabilize the seat during descent) is not deployed;

– Mode 2: moderate speed (250–650 knots or 288–748 mph) and low altitude (<15,000 feet). The pilot parachute is deployed as soon as the seat detaches from its mount. The main parachute is deployed 0.8 to 1 s later;

– Mode 3: high speed (250–650 knots or 288–748 mph) and high altitude (>15,000 feet). The pilot parachute is deployed as soon as the seat detaches from its mount. The system continues to measure the environmental conditions (altitude and speed), delaying the deployment of the main parachute until Mode 2 is attained. The main parachute is then deployed similarly (i.e. after 0.8 to 1 s).

The different stages of Mode 2 are illustrated in Figure 1.3. The real-time constraints associated with each of these stages (① to ⑧) [DOU 88] are summarized in Table 1.1. The firing of the pyrotechnic system of the ejector seat serves as the reference point for the stated time offsets.

One of the essential requirements of the ejection system is maximum protection for the body, limbs and the head. To avoid any injury to the spinal cord, the pilot's body must be held in the correct position, ensuring that the vertebrae are properly arranged (step ③). Ejection is extremely violent: at the moment that the seat is ejected, the pilot experiences acceleration of up to 20G. The pilot's harness is equipped with a torso bracing system that can place the pilot's torso in the correct position for ejection. This system locks, preventing the pilot from moving forward. This system is also subject to strict constraints, as illustrated in Figure 1.4, which shows a block diagram of the harness bracing system patented as early as 1979 [SCH 79]. The harness inflation system that positions the pilot optimally for ejection must be triggered within 40 ms.

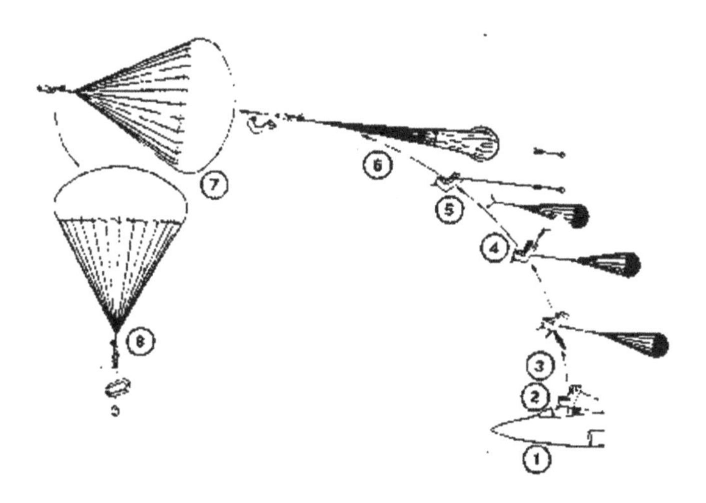

Figure 1.3. *ACES II: operating Mode 2 [DOU 88]*

Event	Time offset (s)
① Rockets fired	0.0
② Stabilizing parachute deployed	0.17
③ Torso bracing system triggered	0.18
④ Main parachute deployed	1.17
⑤ Stabilizing parachute separates from the seat	1.32
⑥ Seat separates from the pilot	1.42
⑦ Main parachute deployed	2.8
⑧ Survival equipment deployed	6.3

Table 1.1. *ACES II: time specifications of the ejection sequence in Mode 2 [DOU 88]*

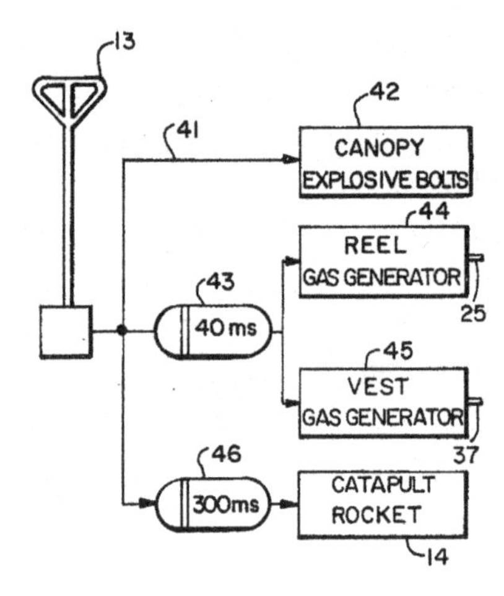

Figure 1.4. *Block diagram of the control system for harness bracing [SCH 79]*

1.3.2. *Multimedia systems*

Multimedia applications have strong time specifications for the storage, transmission and display of video, audio, image and graphic streams. The typical scheduling rate of video streams is between 10 and 20 images per second for videoconferencing, 30 images per second for standard-definition television, 60 images per second for high-definition television (HDTV) and up to 120 images per second for ultra-high-definition television (UHDTV or 4 K) or 8 K.

The necessity of providing ever-increasing rates for these services is a major challenge, and is rendered all the more difficult by the increasing spatial resolution of the image (3840×2160 for 4 K, 7680×4320 for 8K, vs. 1920×1080 currently for HDTV).

Meeting time constraints is more difficult still in interactive applications (live television) for which the compression, transmission and decompression of both streams (video and audio) must satisfy constraints in terms of the end-to-end delay, stream regularity and synchronization. For example, to achieve lip synchronization in a sound/image broadcast, the offset between the audio and video streams must be less than 80 ms. This threshold must be observed to prevent viewers from experiencing discomfort. This system therefore has firm real-time constraints.

For example, consider the real-time transmission sequence of a UHDTV signal, from the camera up until reproduction on the viewer's UHD television (see Figure 1.5).

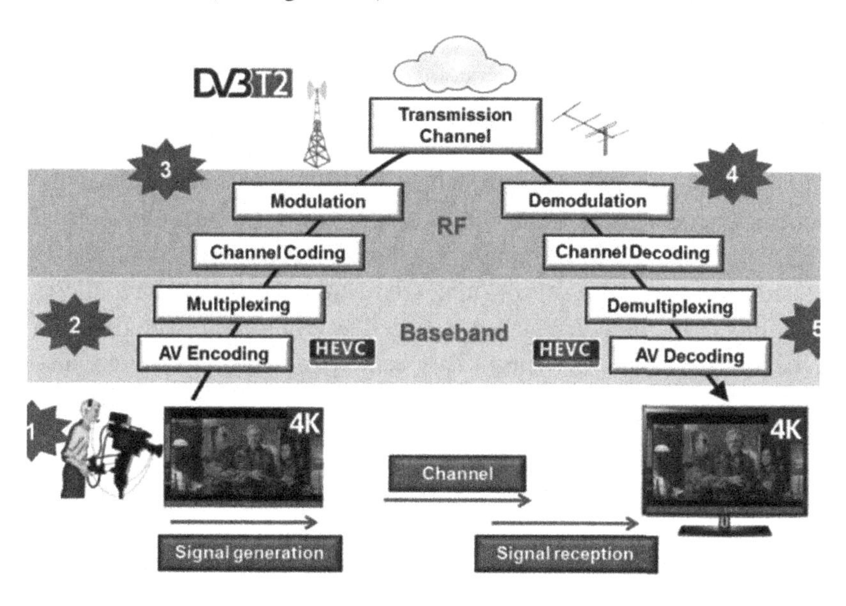

Figure 1.5. *A typical real-time retransmission sequence for UHD signals [DIM 14]*

The UHD camera captures the scene (step 1). In step 2, the signal is sent to the real-time encoder that performs compression according to the HEVC (high-efficiency video coding) standard and multiplexing for subsequent transmission as a transport stream (MPEG-2). In step 3, the UHDTV signal is transmitted over terrestrial connections (DVB-T2), satellite uplinks (DVB-S2), cable networks (DVB-C2) or IP networks (IPTV). In steps 4 and 5, the UHD receiver demodulates and decodes the signal.

Designing an effective real-time software solution for this kind of multimedia application is therefore a major undertaking, especially for the HEVC encoding of the UHD video stream, on which the observance of the real-time constraints of the entire transmission sequence heavily relies.

1.3.3. *Medical systems*

One of the most recent advancements in remote medical monitoring systems is the Diabetes Assistant (DiAs) [KEI 14]. This is a non-invasive continuous blood sugar monitoring device designed to automatically monitor diabetic patients. This device ensures that the blood sugar levels are kept within a narrow bracket around normal levels (i.e. about 70–120 mg/dl). Glucose levels are continuously measured by a glucose sensor implanted below the skin, which allows variations in the blood sugar level as well as hyper- and hypoglycemia to be registered in real time. This remote monitoring system also allows long-distance care by transmitting key information from a smartphone to a remote server via a 3G/wireless connection.

This artificial pancreas is an intelligent network system for monitoring and managing patients' subcutaneous glucose levels. It includes a CGM (continuous glucose monitor), an insulin pump and a *real-time controller*, which are connected via a mobile network (see Figure 1.6).

The real-time controller operates according to a complex algorithm which not only models the physiology of diabetes, but also integrates basic information about each patient (weight, age, lifestyle, blood sugar history), which allows the artificial pancreas to substitute for the patient's bodily function as optimally as possible.

1.4. Real-time operating systems: what are their special features?

The principal difference between a real-time operating system (RTOS) and a general-purpose operating system (GPOS) lies in the fact that the RTOS must be deterministic in the following aspects:

– processing triggers;

– processing duration;

– service and interference of the execution platform.

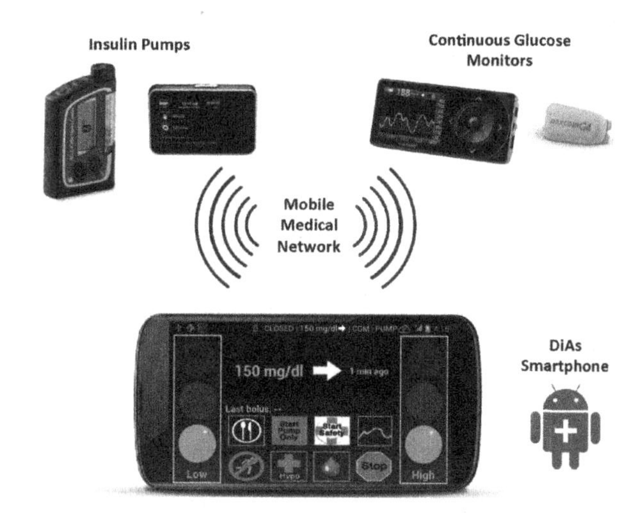

Figure 1.6. *Architecture of the artificial pancreas DiAs [KEI 14]*

The execution time of the services of an RTOS must be fixed, whereas in GPOSes, they can be allowed to vary. In the following, we list the special features offered by RTOSes at the level of individual services.

1.4.1. *Task scheduling*

In multi-tasking systems, multiple processes (or tasks) compete for processor time. The scheduler is the component of the operating system responsible for allocating the processor to each of the various tasks. The scheduler selects the task to execute from the set of ready tasks according to a certain scheduling policy and determines the time that the processor will allocate to this task.

GPOS scheduling strategies are generally designed to optimize the average system performance according to the following objectives:

- *maximal utilization of the processor*;
- *maximal throughput*;
- *minimal cycle time*;
- *minimal waiting time*;
- *minimal response time*.

The throughput is the number of tasks that complete execution per unit time. The cycle time is measured as the time elapsed between the moment at which the task is launched and its completion, including waiting periods. The waiting time, on the other hand, represents the time spent in a waiting queue of ready tasks. Finally, the response time is defined as the time between the sending of a request and the initiation of a system response to the request.

In order to optimize these criteria as much as possible, GPOSes implement scheduling policies that attempt to achieve service fairness for all tasks. This is, for example, the case for the Linux-based *Completely Fair Queuing* (CFQ) scheduler that serves multiple waiting queues by round-robin scheduling [BOV 05]. This type of general-purpose scheduling supports the notion of priority while avoiding situations of process starvation. It is perfectly adapted for systems in which no information on the system load is available beforehand.

The goal of RTOS schedulers is completely different: the objective is to guarantee that all tasks will execute within their respective time constraints, even in worst-case scenarios. Real-time scheduling policies are usually based on priority algorithms (either fixed or dynamic), as we will see in Chapter 2.

1.4.2. *Kernel preemption*

In RTOSes, it is essential that higher priority tasks are able to preempt system calls instead of having to wait for them to terminate, even if these calls were invoked by a lower priority task. For this reason, RTOSes integrate preemption of this type of operation into their internal design. To achieve this, the size of the kernel is generally reduced as much as possible and only includes requests for short services and/or services that are critical for system operation. As a result, kernel services achieve lower latencies, which helps to ensure predictable and bounded response times.

1.4.3. *Dynamic memory allocation*

The fastest and most deterministic approach to memory management is to simply disallow any form of dynamic memory allocation in the programming. Such an approach is, however, not always possible. Having recourse to dynamic memory allocation helps to meet the growing flexibility requirements of applications. Dynamic memory allocation algorithms, however, represent vectors of indeterminism in the domain of real-time systems, due to their unbounded execution times and the heap fragmentation that they generate. The problem of fragmentation is commonly resolved in GPOSes using a garbage collector. However, this is intrinsically not deterministic and so is unsuitable for real-time applications.

Some RTOSes therefore choose to only allow dynamic memory allocation outside of the kernel. Thus, the kernel remains fast and deterministic. The sole problem is that users no longer benefit from memory protection mechanisms outside of the kernel. Another approach for "dynamic" memory allocation is to simply statically allocate fixed-size partitions at compilation, which will be used to fulfill dynamic memory allocation requests at execution. However, the size of these partitions must be carefully chosen so that they can meet the dynamic memory requirements in worst-case scenarios. The final conceivable solution for supporting dynamic memory allocation in real-time systems with strict constraints is to use a memory allocator that performs the operations of allocation and deallocation in constant time $O(1)$, such as the TLSF (Two-Level Segregate Fit) algorithm [MAS 04], which has a maximum observed fragmentation of less than 25%.

1.4.4. *Interrupt handling*

An interrupt is a signal indicating an event that requires an action:

– software interrupt: signal sent by a program (division by zero, stack overflow, etc.);

– hardware interrupt: signal sent by a hardware component (keyboard, mouse, network card, clock, etc.)

When an interrupt is raised, the signal arrives at the processor which then executes a program (function) called the interrupt handler. Interrupts are identified by numbers (which can be shared: multiple sources (peripherals) can generate the same interrupt). Interrupt handlers declare the number of the interrupt that they handle when they register with the kernel. When this interrupt arises, the kernel then calls the corresponding handler.

Firstly, GPOSes do not provide guarantees on the interrupt latency , i.e. the time elapsed between the receipt of the interrupt and its processing. Furthermore, interrupt processing is always given precedence, even at the expense of higher priority application tasks. This execution paradigm is therefore problematical for real-time applications, whose time constraints could be jeopardized by the processing of interrupts at the system level. Consequently, RTOS interrupt handlers are often stripped down to a strict minimum (disruption to real-time tasks preempted by interrupt handling is therefore minimized), and interrupt routines judged to have higher or lower priority than the application are executed in the form of tasks with a fixed priority value, and scheduled for processing together with application tasks.

1.4.5. *Hardware configurations*

There are also differences between GPOSes and RTOSes in terms of the configuration of the hardware on which these operating systems are generally run. Today, personal computers are often equipped with one (or more) powerful processor(s) and large amounts of RAM. Real-time systems, on the other hand, are typically subject to physical constraints such as limited available resources (memory, network, processor, energy). Instead of GPOSes, which are often too "heavy" for these lighter classes of hardware, RTOSes more suitable for embedded systems are used.

1.5. Examples of real-time operating systems for embedded systems

1.5.1. *FreeRTOS*

Available under a GPL license, this open-source RTOS for microcontrollers and small microprocessors is one of the leading operating systems for embedded systems. It is intended for small hardware setups (applications with low memory footprints) in a widely varied range of domains such as health, automobile, home security and medical care. For example, it is used by the Pebble watch and the Netatmo weather station. FreeRTOS is developed and maintained by the British company Real Time Engineers.

FreeRTOS is written in C. In terms of the code structure, the core of FreeRTOS is provided by a few kernel files (task management, semaphores, timers, queues, etc.). These files are shared by all supported architectures. In addition to this, systems include a "portable" folder containing the source files specific to the system (ports, interrupts, etc.). FreeRTOS is equipped with tracing and debugging tools, allowing system operation to be monitored. The ROM memory footprint is located between 6 KB and 10 KB (see Table 1.2 for a summary of the important properties of FreeRTOS). The RAM memory footprint is application-dependent: each task uses about 64 bytes and each queue about 76 bytes.

FreeRTOS offers memory protection mechanisms to manage potential unexpected errors arising from the application level. It also provides tasks with all mechanisms necessary for dynamic memory allocation and stack overflow management at execution. It behaves deterministically and supports three kinds of scheduling: preemptive with fixed priorities, cooperative and round-robin. FreeRTOS further has an integrated mechanism for saving energy in the system. A so-called “idle” task with the lowest kernel priority is responsible for placing the microcontroller on standby when no application tasks are being executed.

Characteristics	Details
Supported systems	> 40: x86, Atmel, TI, Xilinx, Altera, Actel, etc.
Multicore	Not supported
ROM memory footprint	5KB-10KB
Modularity	Configurable, modular
Real-time performance	Deterministic RTOS, context switching in 84 processor cycles
Scheduling	Preemptive priority-based/cooperative/round-robin
Energy management	DPM

Table 1.2. *FreeRTOS: important properties*

Finally, in 2013, FreeRTOS signed a cooperation agreement with the Norwegian organization Nabto with the goal of developing a (commercial) product offering a real-time solution for the Internet of Things (IoT), for example in the context of applications for monitoring and control. The FreeRTOS+Nabto platform (see the architecture illustrated in Figure 1.7) offers a low-bandwidth peer-to-peer connectivity service with the added feature of data encryption. Connected devices running the FreeRTOS+Nabto platform can therefore be accessed and operated either locally or remotely (from a computer, tablet or smartphone) via a rich HCI (human-computer interface).

1.5.2. *μC/OS-III*

Developed by the Canadian company Micrium, the real-time kernel μC/OS-III is the next generation of the well-known kernel μC/OS-II. It can be used for many different types of embedded devices, including those with strict certification constraints such as the medical, aviation or industrial sectors. This RTOS is:

– compliant with the DO-178B standard *"Software Considerations in Airborne Systems and Equipment Certification"* for the regulation of software development in the aviation sector;

– compliant with American market regulations for medical devices according to the 510 (k) procedure defined by the FDA (Food and Drug Administration);

– compliant with the IEC 61508 standard for the safe operation of electrical, electronic or programmable electronic security systems.

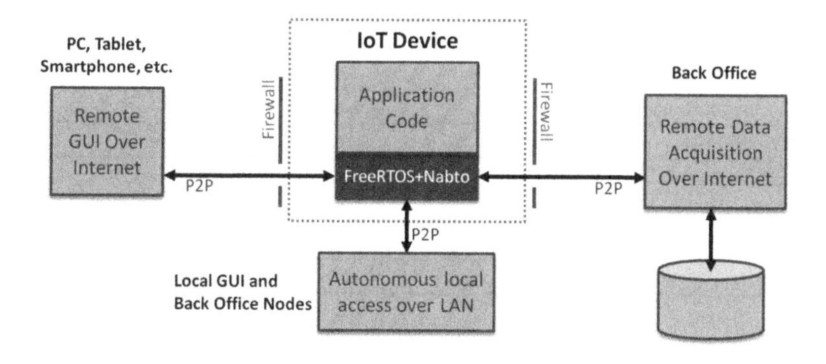

Figure 1.7. *Architecture of the FreeRTOS+Nabto platform for the IoT [FRE 16]*

μC/OS-III may be deployed without license fees or royalties. Its source code, written in C, is open and complies with the ANSI-C (American National Standards Institute) standard. Thanks to its modular structure, its size can be adjusted to only include required functionality and thus is especially suitable for embedded systems with strong memory constraints (maximum memory footprint of 24 KB; see Table 1.3).

In terms of its technical specifications, this deterministic RTOS allows unlimited numbers of tasks, priorities and kernel objects. Two types of scheduling are supported: preemptive priority multitasking and round-robin. It has integrated mechanisms for inter-task communication and synchronization such as semaphores, mutexes and queues. Objects in the RTOS can be allocated either statically or dynamically. The PIP (Priority Inheritance Protocol) is used to avoid priority inversion issues from simultaneous access to shared resources. Dynamic memory allocation is also supported. For interrupt handling, the publisher emphasizes that μC/OS-III achieves very low latency times. Finally, the OS supports advanced functionality for communication that notably includes support for the CAN, TCP/IP, USB and Modbus protocols. Figure 1.8 provides a global overview of the set of controllable modules that the RTOS μC/OS-III can operate.

Characteristics	Details
Supported systems	Altera, Atmel, ARM, Microsoft, IBM, Xilinx, TI, etc.
Multicore	AMP support
ROM memory footprint	6KB-24KB ROM
Modularity	Highly modular
Real-time performance	Deterministic RTOS, real-time guarantees with strict constraints
Scheduling	Preemptive priority-based/round-robin
Energy management	Not supported

Table 1.3. *μC/OS-III: important properties*

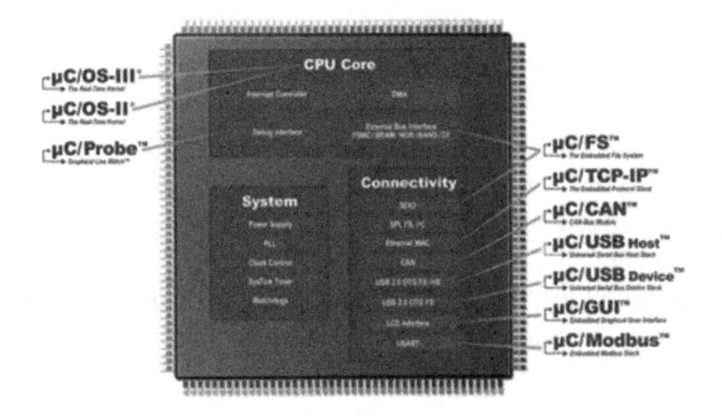

Figure 1.8. *Controllable modules for the μC/OS-III [DIS 16]*

1.5.3. *Keil RTX*

Owned by the British company ARM, Keil RTX is a deterministic RTOS optimized for ARM Cortex-M processors. It has an integrated preemptive priority-based scheduler (254 levels), a cooperative scheduler and a round-robin scheduler. This RTOS guarantees that all operations performed by tasks are *thread-safe* (i.e. functions are implemented in such a way that they can be executed by multiple tasks simultaneously). To achieve this, the kernel implements mutual exclusion to avoid situations of competition and to ensure that when a task attempts to access a shared resource, other tasks are blocked while waiting for the resource. The number of tasks and inter-task synchronization and/or communication objects that can be created is unlimited. The principal features of the real-time kernel Keil RTX are summarized in Table 1.4.

Characteristics	Details
Supported systems	ARM, Cortex-M
Multicore	N/A
ROM memory footprint	4KB
Modularity	Highly modular
Real-time performance	Deterministic RTOS, context switching in less than 300 processor cycles
Scheduling	Preemptive priority-based/cooperative/round-robin
Energy management	DPM

Table 1.4. *Keil RTX: important properties*

One key advantage is that the Norwegian company Energy Micro extended the functionality of the Keil RTX kernel to add a deep standby mode with ultra-low energy consumption that can be toggled between the execution of tasks. Thanks to this extension, the EFM32 series of Cortex-M core microcontrollers can remain in a deep standby state with an energy consumption of less than 1 μA.

To make porting easier, the kernel implements the RTOS API of the ARM standard CMSIS (Cortex Microcontroller Software Interface Standard). Keil RTX is available under a BSD license without fees, including for commercial usage.

1.5.4. *Nucleus*

Nucleus is a stable and optimized RTOS that is currently deployed on more than 3 billion embedded systems [LE 16] for applications in the medical, automobile, security and aerospace sectors. Based on a micro-kernel architecture, in its minimal form, it can be configured to require only 2 KB of memory. It supports many types of hardware: microcontrollers, DSPs, FPGAs and MPUs. An overview of some important properties of Nucleus is given in Table 1.5.

Characteristics	Details
Supported systems	ARM, Atmel, Altera, Samsung, TI, ...
Multicore	AMP and SMP support
ROM memory footprint	2-30KB
Modularity	Highly configurable and modular
Real-time performance	Deterministic RTOS, real-time guarantees with strict constraints, low context-switching time, low interrupt response latency
Scheduling	Preemptive priority-based
Energy management	DPM/DVFS

Table 1.5. *Nucleus: important properties*

This deterministic RTOS is perfectly adapted for real-time systems with strict constraints. The company Mentor Graphics that develops Nucleus emphasizes the rapid OS start-up time and low interrupt response and context-switching latencies. Classical mechanisms of communication and synchronization are available: semaphores, mutexes with PIP support, message queues and mailboxes. Nucleus is compatible with the POSIX and ANSI-C/C++ standards. Scheduling is preemptive priority-based and a multicore version is available for architectures with multicore processors. A number of communication protocols are supported: SPI I^2C, CAN, USB and OTG. The RTOS also includes an array of networking protocols (IPv4/IPv6, TCP/UDP, DHCP client, FTP, Telnet, SSH, etc.) and is certified with the Telecommunications Technology Association Certification (TTA).

One of the key strengths of Nucleus is that it integrates advanced power management functionality (DVFS, deep standby modes and power/clock control), as illustrated in Figure 1.9.

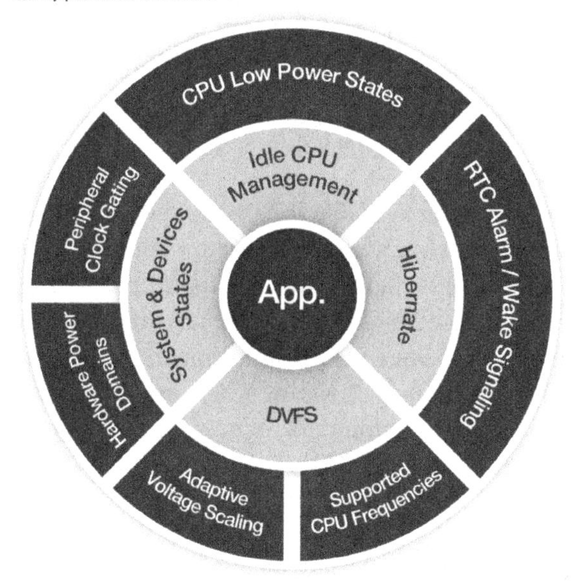

Figure 1.9. *Power management functionality of the RTOS Nucleus [NUC 16]. For a color version of this figure, see www.iste.co.uk/chetto/energyautonomy.zip*

Nucleus is distributed with its source code and may be deployed without license fees or royalties.

1.5.5. *ThreadX-lite*

An RTOS with preemptive priority-based scheduling, ThreadX-lite is adapted for development on ARM Cortex-M cores (microcontrollers Cortex-M0, M0+, M3 and M4-based microcontrollers marketed by Atmel, Freescale, Fujitsu, Infineon, NXP, STMicroelectronics and Texas Instruments). Table 1.6 summarizes its principal characteristics.

Characteristics	Details
Supported systems	ARM Cortex-M
Multicore	SMP and AMP support
ROM memory footprint	2 KB
Modularity	Highly modular
Real-time performance	Deterministic pico-kernel, real-time guarantees with strict constraints, context switching in 100 processor cycles
Scheduling	Preemptive priority-based/cooperative
Power management	Not supported

Table 1.6. *ThreadX-lite: important properties*

This RTOS was made available in 2012 as the fruit of a partnership between Express Logic (software publisher for embedded systems) and IAR Systems (Swedish company specializing in the development and debugging of embedded software). ThreadX-lite has the classical services offered by RTOSes (task management, queues, semaphores, timers, etc.). It requires only 2 KB of ROM memory and less than 1 KB of RAM memory. According to Express Logic, context switching on a Cortex-M3 running at 100 MHz requires less than 1 ms.

ThreadX-lite is only available in a packaged commercial version distributed by IAR Systems. Deployment of ThreadX-lite is, however, royalty-free.

1.5.6. *Contiki*

Contiki is an open operating system for microcontrollers and embedded systems, primarily dedicated to wireless mini-sensors. Developed in 2004 at the Swedish Institute of Computer Science, it is available for free under a BSD license (the source code may be reused in full or in part without restrictions).

Contiki is ultralight (less than 10 KB of RAM and 30 KB of ROM) with very low electrical energy consumption. It is supported by a wide range of platforms, including apple2enh, atari, C128, C64, Sentilla-usb, msb430, ESB, etc. The main characteristics of Contiki are summarized in Table 1.7.

Characteristics	Details
Supported systems	apple2enh, atari, C128, C64, msb430, ESB, etc.
Multicore	Not supported
ROM memory footprint	30KB
Modularity	Highly modular
Real-time performance	Soft
Scheduling	Preemptive priority-based
Energy management	Very low consumption

Table 1.7. *Contiki: important properties*

Written in C for improved portability, this RTOS is modular and supports dynamic module loading. With its low resource requirements, Contiki is also designed as an OS for the IoT. One of its strengths is its simplified development environment based on a virtual machine.

Contiki uses a preemptive priority-based, *time-triggered* scheduler. This mode of scheduling is adapted to the utilization of sensors in which a new processing task is often triggered by the occurrence of an external event. It is also more energy efficient.

1.6. Conclusion

In this chapter, an outline of real-time systems was given, including a description of the different classes to which they can belong (*hard, soft, firm*). We have seen that they can respond to two activation paradigms (*time-based* and *event-based*) according to the nature of the environment and the completeness of the system knowledge of the characteristics of this environment. The features of the very particular execution platforms (RTOSes) used in real-time contexts were also highlighted. Among the important characteristics of RTOSes, we note the capacity to provide integrated guarantees of time-determinism at all levels (task management, memory allocation/deallocation and interrupt handling). In the few listed examples of real-time embedded operating systems, it should be pointed out that only a few have mechanisms for energy management. Those that do are intended for energy efficiency and not energy autonomy, which is a specific feature required, for example, by the recent family of real-time wireless sensor systems. In

Chapter 4, we will present new techniques for efficiently exploiting the processor with the goal of achieving the level of autonomy increasingly demanded by real-time systems as a result of the expansion of the IoT.

We also listed some of the important properties of a few leading examples of RTOSes in the field of embedded systems. Besides the time guarantees provided by these RTOSes, we saw that other non-functional criteria such as the optimization of energy consumption are currently the center of focus. This aspect will be discussed in detail in Chapter 4. As a preliminary to that chapter, the next chapter will discuss the question of real-time scheduling. This synthetic overview of the state-of-the-art in time-constrained systems will prepare the reader for the discussion of a context with two simultaneous real-time constraints: time and energy.

2

Principles of Real-time Scheduling

This chapter gives an introduction to the fundamental principles of real-time scheduling. In the first section, we will discuss the manner in which the temporal characteristics of a real-time system are formalized by giving a precise characterization of application tasks. The second section will introduce the basic approaches and definitions associated with the question of time validation based on the analysis of system schedulability. We will then consider the basic concepts of uniprocessor real-time scheduling and recall the most important scientific results in this area (classification, properties, etc.). The subsequent section will describe in more detail the state-of-the-art in scheduling algorithms for periodic tasks, specifying their schedulability and optimality conditions. Finally, we will briefly present the main types of existing aperiodic task server (servers with static or dynamic priorities).

This chapter is thus organized as follows:

- characterization and models of real-time tasks;
- schedulability analysis;
- uniprocessor scheduling;
- periodic task scheduling;
- aperiodic task servers.

2.1. Characterization and models of real-time tasks

2.1.1. *Definition*

We call a *task* any software entity that performs a particular function within a software application. Each task corresponds to the execution of a given sequence of operations on the processor. Depending on the nature of the service provided by the task, this sequence of operations may be repeated multiple times. Each execution can then be considered individually as a *job* (or *instance*) .

In a multi-tasking environment, tasks can occupy one of the following four states [DOR 91]: *executing*, *ready*, *suspended* or *dormant*. The first three states are considered active states (the tasks exist and fulfill a certain service for the application), whereas the last state is viewed as inactive (such tasks do not exist or no longer exist from the point of view of the application). Transitions from one state to another are decided by the scheduler. In practice, a change of state often results in *context switching*. Figure 2.1 illustrates the active states and transitions between states.

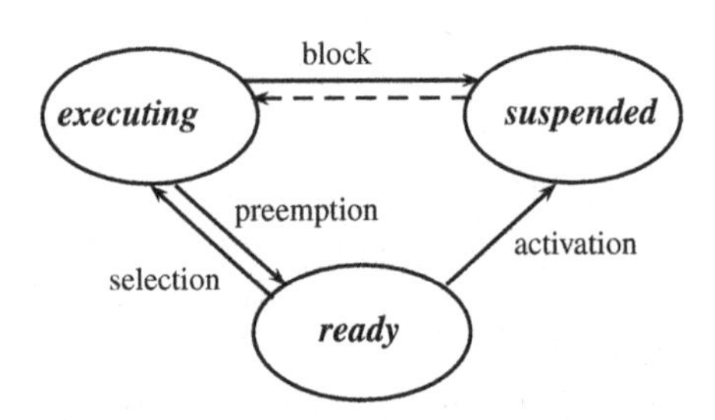

Figure 2.1. *Diagram of active task states*

Tasks in the *executing* state have control of the processor and are currently executing their code. The executing task is the task that is considered to have the highest priority among the candidates for processor allocation at a certain moment in time. *Suspended* tasks are not candidates for processor allocation. The execution of these tasks is temporarily suspended until the task obtains the required resource (other than processor time) for execution. *Ready* tasks are waiting to be selected for execution. *Dormant* tasks have either not yet been created or have been permanently terminated.

A *real-time task* is a task with a critical deadline, i.e. a task with a bound on the time at which its execution must end. The set of tasks to be executed on a given device is called the *task set*. If all tasks in this set are characterized by a deadline, it is described as a critical task set (or a critical-deadline task set).

Other types of constraint (other than time constraints) can of course be placed on the execution of real-time tasks. These can include [SIL 93]:

– *resource constraints*, i.e. constraints arising from the mutual exclusion of access to critical resources. The resources required by a task in the process of being executed are not always assumed to be available when the task requests execution;

– *synchronization constraints*, which may be described by a set of precedence relations (or anteriorities) determining the order in which the tasks must be processed. If there is no precedence relation between two tasks, the tasks are described as independent;

– *execution constraints* based on two modes of task execution, respectively described as *preemptive* and *non-preemptive*. If all tasks are preemptible, their execution may be interrupted at any moment and resumed at a later point (unlike the non-preemptive case, in which each task reserves the processor from the beginning to the end of execution);

– *placement constraints* relating to the identity of the processor(s) in a multi-processor or distributed system that are authorized to execute a task.

Furthermore, three types of task are commonly found: periodic tasks, sporadic tasks and aperiodic tasks. A description of the canonical model of each of these tasks is given below.

2.1.2. *Models of tasks*

A *periodic* real-time task $\tau_i(\phi_i, C_i, T_i, D_i)$ is defined by [COT 00]:

– its activation offset ϕ_i;

– its maximum or worst-case execution time C_i;

– its activation period T_i;

– its critical delay D_i.

The activation offset ϕ_i corresponds to the time at which the first job of τ_i becomes active. The worst-case execution time C_i is the maximal time interval between the beginning and the end of execution (excluding interrupts) on a given processor. The activation period T_i represents the time interval between the activation of two consecutive jobs belonging to the task. The critical delay D_i (or *relative deadline*) corresponds to the time interval between the moment of activation and the deadline. The absolute deadline d_i (also called the critical time) is the latest possible execution end time for the current job of the task. Finally, we define $L_i = D_i - C_i$ as the *static laxity* of the task τ_i, i.e. the maximum possible delay before the start of execution such that the deadline can still be met.

Each task generates a new job J_i^k at each time $r_{i,k} = \phi_i + kT_i$, where k is an integer, such that $k \geq 0$. Each of these jobs must be executed before its absolute deadline $d_{i,k} = r_{i,k} + D_i$. The moment at which the job J_i^k begins (or ends) execution is called the *execution start time* (or *execution end time*), which is written as $s_{i,k}$ (or $f_{i,k}$). We define the *response time* $R_{i,k}$ as the interval between the activation of J_i^k and the end of its execution, so that $R_{i,k} = f_{i,k} - r_{i,k}$.

Figure 2.2 shows two consecutive jobs of a periodic task. Time is displayed along the horizontal axis, whereas the vertical axis indicates whether the task is active (up state) or inactive (down state).

If the earliest activation time of the task is known (or not), the task is described as *concrete* (or *non-concrete*). Furthermore, tasks have:

– *constrained deadlines* if their critical delay cannot exceed their period ($D_i \leq T_i$);

– *implicit deadlines* if the critical delay of the task is equal to its period ($D_i = T_i$). Each job of the task must terminate before the next job of the same task arrives. This is therefore a special case of a constrained-deadline system;

– *arbitrary deadlines* whenever there is no constraint on the critical delay and the period.

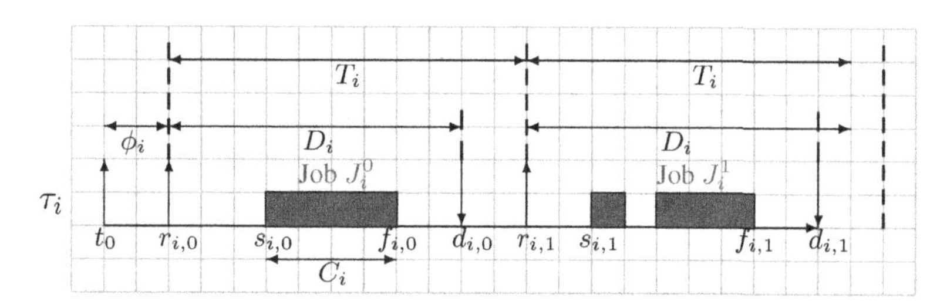

Figure 2.2. *Model of a periodic task* τ_i

Sporadic tasks (a term introduced by Mok [MOK 83], although the concept was familiar beforehand) are similar to periodic tasks except that T_i corresponds to the *minimum* interval (and not the exact interval) between two consecutive activations of jobs belonging to the task τ_i (see Figure 2.3).

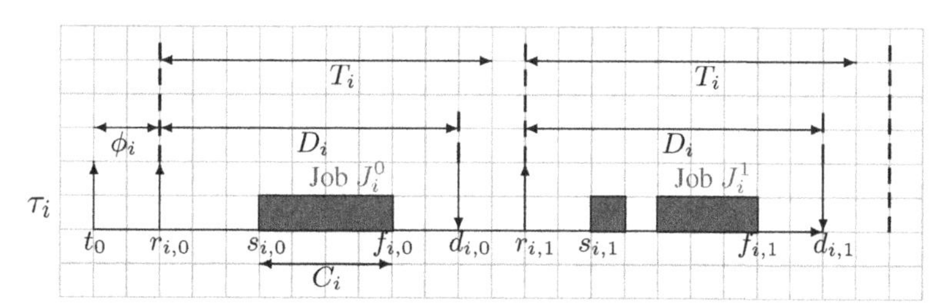

Figure 2.3. *Model of a sporadic task* τ_i

There also exist non-recurring tasks described as *aperiodic*. The moment of activation of these tasks is not known beforehand, and so they do not have a period. The jobs of this kind of task are activated by the occurrence of events that can be either external, if triggered by the environment, or internal, if originating from another task. Within this family of tasks, a distinction is generally made between *non-critical* aperiodic tasks (tasks without a deadline that must be executed at the earliest opportunity) and *critical* aperiodic tasks (tasks endowed with a deadline whose execution must observe this time constraint). The latter category of tasks is illustrated in Figure 2.4.

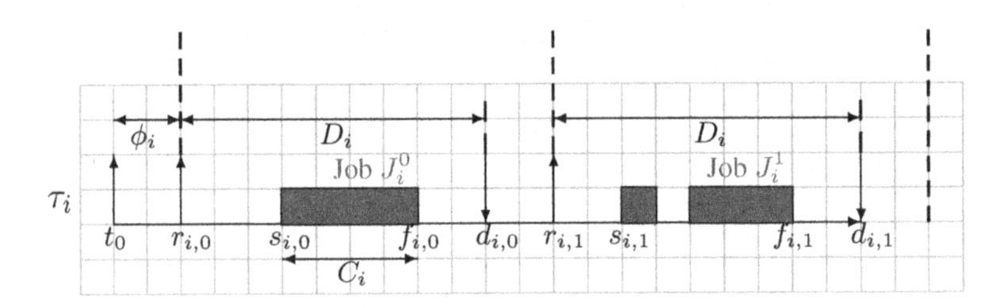

Figure 2.4. *Model of a critical aperiodic task* τ_i

We write $\mathcal{T} = \{\tau_i(\phi_i, C_i, T_i, D_i),\ i = 1..n\}$ for a set of periodic tasks, where n is the number of tasks in the set $\mathcal{T}$. In the context of concrete tasks, a set of periodic tasks is *synchronous* if all tasks in the set have the same activation time, i.e. if $\forall i, j \in \{1...n\}$, $\phi_i = \phi_j$. Otherwise, it is described as *asynchronous*.

2.2. Schedulability analysis

Schedulability analysis allows the verification of time constraints associated with tasks before the application is run.

DEFINITION 2.1.– A schedulability test determines prior to execution whether the deadlines of a given set of tasks will always be observed.

Schedulability tests can be:

– *sufficient*: if this property is present, then the task set is schedulable. If absent, the test cannot be used to conclude that the set of tasks is effectively schedulable;

– *necessary*: if this property is not present, then the corresponding set of tasks is not schedulable. If not, it is nevertheless not possible to conclude that the set of tasks can be scheduled so that all time constraints are met;

– *exact*: the test is both *necessary* and *sufficient*. The test can be used to determine prior to execution whether or not a task set is schedulable.

Exact schedulability tests are the most desirable for time-validating real-time systems. However, this kind of test does not always exist, can be too costly to evaluate or can impose restrictive assumptions on the tasks. Real-time software designers therefore often choose to use tests that are sufficient but not necessary.

There are three techniques of worst-case analysis in the literature [RIC 05]: *utilization factor analysis*, *processor demand analysis* and *response time analysis*. Before exploring these techniques in more detail, we will begin by introducing a few definitions and key concepts associated with schedulability tests.

2.2.1. *Basic definitions associated with schedulability analysis*

2.2.1.1. *The utilization factor*

The *utilization factor* u_i of a task τ_i is defined as the ratio of its execution time over its activation period :

$$u_i = \frac{C_i}{T_i} \qquad [2.1]$$

In general, the *effective processor utilization* of a task set $\mathcal{T}$, denoted U, is defined as the sum over all tasks of the individual utilization factors of each task [LIU 73]:

$$U = \sum_{\tau_i \in \mathcal{T}} u_i \qquad [2.2]$$

Similarly, the maximum utilization factor of a task set $\mathcal{T}$ is denoted as U_{max}:

$$U_{max} = \max_{\tau_i \in \mathcal{T}} u_i \qquad [2.3]$$

2.2.1.2. *The load factor*

The *load factor*, or *density factor*, δ_i of a task τ_i is defined as the ratio of its execution time over the minimum of its activation period and its critical delay :

$$\delta_i = \frac{C_i}{\min(T_i, D_i)} \qquad [2.4]$$

Analogously to the effective processor utilization, we introduce the notion of *effective processor load* (and *maximum processor load*) of a task set $\mathcal{T}$, written δ (or δ_{max}), as the sum (or the maximum) of the individual load factors of each task [LIU 73]:

$$\delta = \sum_{\tau_i \in \mathcal{T}} \delta_i; \qquad \delta_{max} = \max_{\tau_i \in \mathcal{T}} \delta_i \qquad [2.5]$$

2.2.1.3. *Processor demand*

The analysis of processor activity also plays a crucial role in the study of the schedulability of systems with sporadic real-time tasks. In this approach, the number of activated jobs is counted over a certain period of activity. A *busy period* is any time interval during which the processor is continuously occupied by the execution of tasks (i.e. no idle time is observed). In the synchronous case, this interval is defined as:

DEFINITION 2.2.– A synchronous busy period of a processor is an interval of processor activity beginning with the simultaneous activation of all tasks and ending with the first idle period (possibly none).

The *demand bound function* of a task τ_i on a time interval $[t_1, t_2)$, written as $dbf(\tau_i, t_1, t_2)$, is defined as the cumulative duration of the jobs belonging to this task whose activation times and deadlines are contained in the time interval $[t_1, t_2)$ [BAR 90] :

$$dbf(\tau_i, t_1, t_2) = max\left(0, \lfloor\frac{t_2 - \phi_i - D_i}{T_i}\rfloor - \lceil\frac{t_1 - \phi_i}{T_i}\rceil + 1\right) C_i \quad [2.6]$$

Given a task set $\mathcal{T}$, the *demand bound function* of $\mathcal{T}$ on the time interval $[t_1, t_2)$, written as $dbf_{\mathcal{T}}(t_1, t_2)$, is given by:

$$dbf_{\mathcal{T}}(t_1, t_2) = \sum_{\tau_i \in \mathcal{T}} dbf(\tau_i, t_1, t_2) \quad [2.7]$$

2.2.2. *Approaches to schedulability analysis*

In the following sections, we will explore each of the various existing approaches to schedulability analysis in turn. In particular, we will focus on how they differ by either considering all tasks simultaneously to determine the schedulability of a task set or simply considering the busy periods of the processor.

2.2.2.1. *Utilization factor analysis*

Utilization factor analysis establishes a worst-case *performance guarantee*. The effective processor utilization is calculated for a given task set to verify whether it exceeds the threshold specific to the considered scheduling algorithm.

2.2.2.2. *Processor demand analysis*

Processor demand analysis checks that the cumulative sum of the processing times associated with all tasks over any time interval does not exceed the processor capacity. In other words, this approach tests that on any time interval L, the cumulative maximum duration (or an upper bound thereof) of the execution of jobs for tasks whose activation and deadline occur in the interval does not exceed the size of the interval (so is less than or equal to L).

2.2.2.3. *Response time analysis*

This two-step approach calculates the worst-case response time R_i (or an upper bound thereof) of each task and checks that it does not exceed the critical delay (i.e. $\forall i, 1 \leq i \leq n, R_i \leq D_i$). A task is schedulable if and only if its worst-case response time is less than or equal to its critical delay. In the context of fixed-priority scheduling, this method is often described as RTA (Response Time Analysis) [JOS 86].

In the following section, we will introduce the topic of real-time scheduling using the task models described in section 2.1.2.

2.3. Uniprocessor scheduling

One of the challenges associated with scheduling real-time tasks is resolving processor access conflicts between tasks as a result of simultaneous access requests according to the relative urgency of the tasks. This scheduling function is provided by a service procedure belonging to the operating system (the *scheduler*), whose role is to allocate the processor to each of the different tasks. Given resource constraints in terms of synchronization, execution and timing, and according to the performance criteria of the application, the scheduler must determine which task to execute at any given moment in time. The scheduler can implement one or multiple *scheduling algorithms* which specify a policy for allocating the processor to tasks. The timetable of tasks constructed by a scheduling algorithm is called the *schedule* produced by this algorithm for the given task set.

2.3.1. *Classification of scheduling algorithms*

Scheduling algorithms are usually classified according to the characteristics of the system on which they are implemented. This leads to the following classification:

– Uniprocessor or multiprocessor

If all tasks are only executed on one single processor, scheduling is described as *uniprocessor*. If not, the term *multiprocessor* scheduling is used to indicate that the system possesses multiple available processors.

– Idling or non-idling

In the case of *non-idling* scheduling, ready tasks are not allowed to enter a suspended state if processor resources are free. We say that the scheduler runs without inserted idle time. In the case of *idling* scheduling, if a task is ready, it may either be selected or suspended for a certain period of time before being activated, even if the processor is available.

Idling scheduling can be particularly useful in non-preemptive contexts. Indeed, some task sets that are unschedulable in non-preemptive non-idling systems are schedulable in non-preemptive idling systems [DEC 02].

– Optimal or non-optimal

By definition, a scheduling algorithm is said to be *optimal* [GEO 96] for a given class of scheduling problems if it can correctly schedule a set of tasks whenever this set is feasible (a set is feasible whenever the execution of tasks can be arranged in such a way that all deadlines are met). Consequently, if a set is not schedulable by an optimal algorithm for a class containing the set, then it is not schedulable by any algorithm in this class.

– Online or offline

The simplest method is to plan the scheduling *offline*. At execution, the tasks are then executed according to the planned schedule. Offline scheduling is particularly suitable for situations where the majority of activities are called periodically. However, this is generally not the case in real-time contexts. Static cyclic scheduling is not suitable whenever events are liable to occur aperiodically, in which case online schedulers are preferred, which determine the schedule during system execution. Online execution is more flexible in the sense that it allows fluctuating parameters to be taken into account during execution (jobs ending earlier, arrival of a new task, etc.). *Online* scheduling must always be complemented by offline schedulability analysis to ensure that time constraints will be observed at execution.

– Preemptive or non-preemptive

A necessary condition for a scheduler to be *preemptive* is for all tasks to be preemptible. In this case, if a task is judged more urgent or higher priority, it can gain control of the processor, at the expense of another task whose execution is interrupted and resumed at a later point. Note that in the case of non-preemptive uniprocessor scheduling, managing critical resources is no longer an issue, as simultaneous access to resources cannot exist without preemption.

– Clairvoyant versus non-clairvoyant

A scheduling algorithm is said to be *clairvoyant* if it has full knowledge of the set of future tasks and their constraints (deadlines, execution times, precedence constraints, future activation times, etc.). By contrast, it is described as *non-clairvoyant* if it only knows the arrival date of a task at the moment at which it arrives.

– Centralized or distributed

Scheduling is *distributed* if scheduling decisions are made by an algorithm locally at each node. It is *centralized* if the scheduling algorithm responsible for the whole system operates from a single central node, whether or not the system itself is distributed.

2.3.2. *Properties of scheduling algorithms*

In this section, we present a description of the properties of scheduling algorithms in terms of the analysis of *schedulability* and *feasibility*:

DEFINITION 2.3.– A valid schedule produced by a scheduling algorithm for a given task set is a schedule in which the constraints of each of the tasks are met.

DEFINITION 2.4.– Scheduling is feasible for a given task set if there exists at least one scheduler capable of producing a valid schedule.

DEFINITION 2.5.– A task set is guaranteed schedulable by a scheduling algorithm if and only if the schedule produced by this algorithm is valid.

DEFINITION 2.6.– A task set is schedulable if there exists an algorithm for which it is guaranteed schedulable.

DEFINITION 2.7.– A scheduling algorithm is said to be optimal for a class of systems and a set of scheduling policies given certain assumptions if and only if any system schedulable by some policy in this set is guaranteed schedulable by this algorithm.

2.3.3. *Scheduling algorithm complexity*

As a complement to the metrics described above, the effectiveness of an algorithm is also evaluated as a function of its computational *complexity* [AHO 74, HOR 76]. In general, the complexity is calculated by counting the number of elementary operations used, i.e. the number of basic instructions in any programming language (addition, subtraction, assignment, comparison, etc.), as a function of the number of problem inputs.

Let f be the complexity function of an algorithm representing the largest number of elementary operations required by the algorithm to solve a problem Π, and let n be the size of the problem Π, corresponding to the number of input data points required to describe Π. This leads to the following definitions [BOU 91]:

DEFINITION 2.8.– An algorithm is said to run in polynomial time if its complexity function f is $O(p(n))$, where p is a polynomial. If p is linear, the algorithm is said to have linear complexity.

DEFINITION 2.9.– An algorithm is said to run in pseudo-polynomial time if its complexity function f takes the form of a polynomial function but not in the size of the problem. The execution time depends not only on the value of the problem inputs but also their lengths.

DEFINITION 2.10.– An algorithm is said to run in exponential time if its complexity function f is $O(n!)$ or $O(k^n)$ for $k \geq 1$, or alternatively $O(n^{\log n})$.

Polynomial-time algorithms are of course the most desirable as they provide solutions to scheduling problems in a timespan that is considered to be reasonable, unlike exponential-complexity algorithms. Note that the higher the algorithmic complexity of a scheduler, the more expensive its implementation overhead.

2.4. Periodic task scheduling

The majority of real-time dynamic schedulers are based on the notion of *priority*. If the priorities of all tasks are fixed at initialization and do not change over the lifetime of the application, the algorithm is said to be *fixed-priority*. If priorities change over time, the algorithm is described as *dynamic-priority*. In the following sections, we present the most commonly encountered priority-based algorithms. For each algorithm, we give a description of its behavior as well as its performance and associated schedulability conditions.

First, we recall a fundamental result on preemptive scheduling, established by Leung and Merril [LEU 80]:

THEOREM 2.1.– The schedule produced by any preemptive algorithm for a set of periodic tasks is always periodic with period equal to the least common multiple (LCM) of the periods of the tasks in the set, denoted by H.

This result allows us to deduce that if the processor is idle at a given time t, it will also be idle at time $t + kH$ (for $k \in \mathbb{N}$). Similarly, if it performed x_i units of work on a job of τ_i at moment t, then it will also have performed τ_i units of work on another job of τ_i at time $t + kH$.

This result also allows us to reduce the schedulability analysis to a reference window, called the hyperperiod, equal to the LCM of the periods.

2.4.1. *Fixed-priority scheduling*

2.4.1.1. *Rate-monotonic scheduling (RMS)*

2.4.1.1.1. Description

The most commonly used scheduling algorithm is the *rate-monotonic scheduling (RMS)* algorithm. In this algorithm, the priority of a task is inversely proportional to its activation period (conflicts are resolved arbitrarily) [LIU 73]. In other words, the smaller the activation period T_i of a task, the higher its priority. Consider a set of tasks $\mathcal{T} = \{\tau_i(C_i, T_i, D_i), i = 1..3\}$ scheduled by RMS (see Figure 2.5). Assume that all tasks have implicit deadlines (i.e. $D_i = T_i$). Assume further that they are synchronous with initial activation offset $\phi_i = 0$.

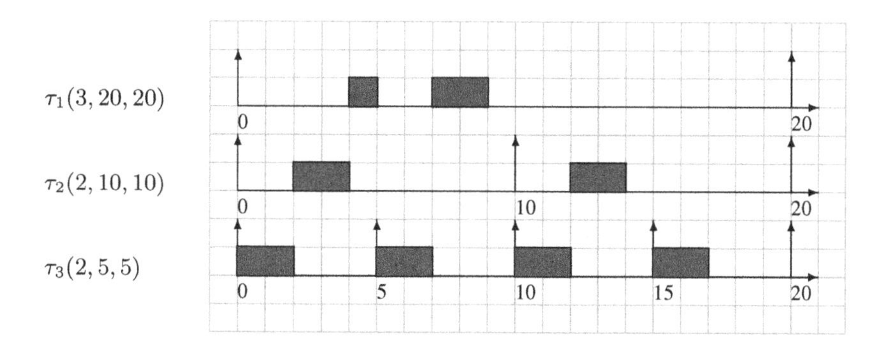

Figure 2.5. *Illustration of the working principle of RMS*

The task with the smallest period (τ_3 in the example) has the highest priority. It is executed immediately at the start of its period of execution, if necessary by interrupting lower-priority tasks (τ_1 is interrupted at time $t = 5$ in favor of τ_3). The benefit of this scheduling strategy lies in the simplicity of its implementation, requiring only one single list structure in order for the algorithm to function.

2.4.1.1.2. Optimality results

RMS is optimal in the class of preemptive fixed-priority algorithms for independent periodic tasks with implicit deadlines [LIU 73]. This means that if a task set $\mathcal{T}$ is guaranteed schedulable by some preemptive fixed-priority algorithm, then it is guaranteed schedulable by RMS.

RMS is not optimal in a non-preemptive context, or if the assumption $\forall i = 1..n$, $D_i = T_i$ does not hold.

2.4.1.1.3. Schedulability conditions

A sufficient condition for schedulability was formulated by Liu and Layland [LIU 73] for the RMS algorithm:

THEOREM 2.2.– Any set of n periodic tasks with implicit deadlines is guaranteed schedulable by RMS if its effective processor utilization U

satisfies:

$$U = \sum_{i=1}^{n} \frac{C_i}{T_i} \leq n(2^{\frac{1}{n}} - 1) \qquad [2.8]$$

We can therefore deduce that as the number of tasks tends to infinity, the minimum utilization factor guaranteeing that the set is schedulable approaches $\log 2$, which is approximately 69.3%. Note, however, that for RMS, the schedulability bound is higher if the periods of the tasks are harmonic, i.e. are multiples of other periods.

Much work has been done to improve the schedulability bound of the RMS algorithm and to relax some of the restrictive hypotheses on the set of considered tasks. Lehoczky *et al.* [LEH 89] conducted a statistical study showing that for sets of tasks with randomly generated parameters, task sets with an effective processor utilization of approximately 88% are guaranteed schedulable by the RMS algorithm (this is, however, only a statistical result to demonstrate the pessimism of the test in theorem 2.2). Also, a schedulability test of equivalent complexity to the test formulated by Liu and Layland, known as the *Hyperbolic Bound (HB)* test, was suggested by Bini *et al.* in 2001 [BIN 01]. This test, which states that *a set of periodic tasks is schedulable by RMS if* $\prod_{i=1}^{n} (U_i + 1) \leq 2$, improves the schedulability ratio by a factor of up to $\sqrt{2}$. In the general case, exact schedulability tests leading to necessary and sufficient conditions were independently derived by Joseph and Pandya [JOS 86], Lehoczky *et al.* [LEH 89] and Audsley *et al.* [AUD 93]. The exact test derived by Lehoczky *et al.* [LEH 89], based on *processor demand analysis*, is given below:

THEOREM 2.3.– Let $\mathcal{T}$ be a set of n independent periodic tasks with implicit deadlines, such that $T_1 \leq T_2 ... \leq T_n$. $\mathcal{T}$ is schedulable by RMS if and only if:

$$\max_{i=1..n} \{\min_{t \in S_i} \sum_{j=1}^{n} \frac{C_j}{t} \lceil \frac{t}{T_i} \rceil\} \leq 1 \qquad [2.9]$$

where $S_i = \{D_i\} \cup \{kT_j, 1 \leq j \leq i, k = 1, ..., \lfloor \frac{T_i}{T_j} \rfloor\}$

In terms of the complexity of the algorithm, the number of iterations is bounded by the ratio $\frac{T_i}{T_j}$, which implies that the test has pseudo-polynomial complexity.

2.4.1.2. *Deadline-monotonic scheduling (DMS)*

2.4.1.2.1. Description

In the context of *deadline-monotonic scheduling (DMS)*, the priority of each task is inversely proportional to its relative deadline (conflicts are resolved arbitrarily) [LEU 82]. In other words, the smaller the critical delay D_i of a task, the higher its priority. The DMS algorithm is illustrated in Figure 2.6 with a set of tasks $\mathcal{T} = \{\tau_i(C_i, T_i, D_i), i = 1..3\}$. The tasks are assumed to be synchronous with initial activation offset $\phi_i = 0$.

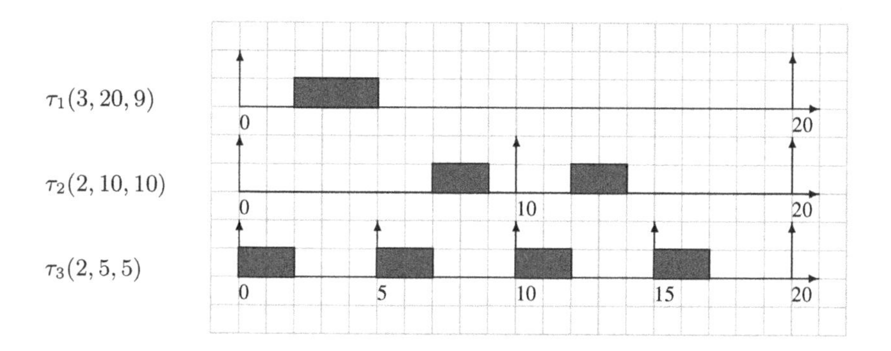

Figure 2.6. *Illustration of the working principle of DMS*

2.4.1.2.2. Optimality results

DMS is optimal in the class of non-idling preemptive fixed-priority algorithms for independent periodic tasks, such that $D_i \leq T_i$ [LEU 82, AUD 91, BUT 97]. This means that, given these assumptions, if a task set $\mathcal{T}$ is guaranteed schedulable by some preemptive fixed-priority algorithm, then it is also guaranteed schedulable by DMS.

DMS is only optimal in non-preemptive contexts if $\forall(i,j)$, $C_i \leq C_j \Rightarrow D_i \leq D_j$ [BAT 98].

2.4.1.2.3. Schedulability conditions

In the case of the DMS algorithm, schedulability may be checked using the RMS schedulability test after replacing the utilization factor by the load factor:

THEOREM 2.4.– Any set of n periodic tasks with implicit deadlines is guaranteed schedulable by DMS if its load factor δ satisfies:

$$\delta = \sum_{i=1}^{n} \frac{C_i}{D_i} \leq n(2^{\frac{1}{n}} - 1) \qquad [2.10]$$

Thus, for DMS, we also have an upper bound ($\approx$ 69.3%) for the effective processor utilization as $n \to \infty$.

Feasibility conditions for the DMS algorithm were also developed by Audsley and Burns [AUD 90], including a *necessary and sufficient* feasibility test based on an algorithm of *data-dependent* complexity. This test is capable of determining the feasibility of any set of fixed-priority tasks, such that $D_i \leq T_i$, regardless of the chosen priority assignment criteria. Later, Lehoczky *et al.* [LEH 89] extended the above-stated test for RMS (see theorem 2.3) based on processor demand analysis to the DMS algorithm (also with pseudo-polynomial complexity):

THEOREM 2.5.– Let $\mathcal{T}$ be a set of n independent periodic tasks with constrained deadlines, such that $D_1 \leq D_2 ... \leq D_n$. $\mathcal{T}$ is schedulable by DMS if and only if:

$$\max_{i=1..n} \{\min_{t \in S_i} \sum_{j=1}^{n} \frac{C_j}{t} \lceil \frac{t}{T_i} \rceil\} \leq 1 \qquad [2.11]$$

where $S_i = \{D_i\} \cup \{kT_j, 1 \leq j \leq i, k = 1, ..., \lfloor \frac{D_i}{T_j} \rfloor\}$

2.4.2. *Dynamic-priority scheduling*

2.4.2.1. *Earliest Deadline First*

2.4.2.1.1. Description

By definition, the Earliest Deadline First (EDF) algorithm gives priority at any given moment in time to the task with the earliest absolute deadline d_i [JAC 55, SER 72, LIU 73]. In the event of a conflict, the task with the earliest activation time can be executed first. The EDF algorithm is illustrated in Figure 4.3 with a set of tasks $\mathcal{T} = \{\tau_i(C_i, T_i, D_i), i = 1..3\}$.

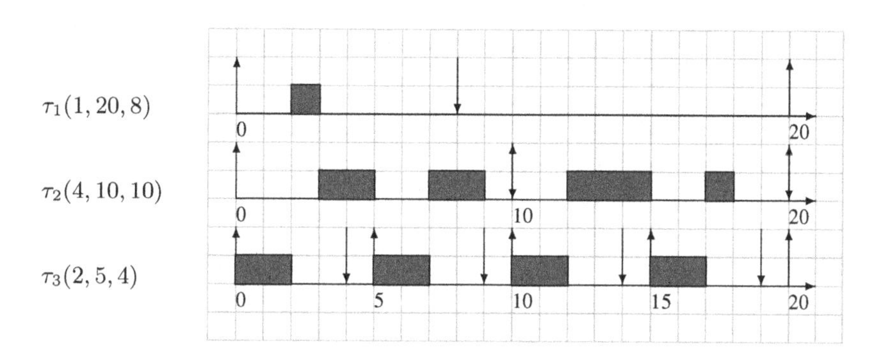

Figure 2.7. *Illustration of the working principle of EDF*

2.4.2.1.2. Optimality results

EDF is optimal in the class of preemptive algorithms for independent periodic tasks [DER 74]. If a set of independent periodic tasks is guaranteed schedulable by some algorithm, then it is also guaranteed schedulable by EDF.

In the non-preemptive case, the scheduling problem is well known for being NP-hard [RID 05]. However, for non-idling schedulers, the problem once again becomes solvable. The EDF algorithm is optimal in the sub-class of non-preemptive schedulers, as was shown by George *et al.* in [GEO 95].

2.4.2.1.3. Schedulability conditions

Given the assumption of synchronous tasks with implicit deadlines ($D_i = T_i$), the following schedulability test can be established for the EDF algorithm based on a *necessary and sufficient* condition [LIU 73]:

THEOREM 2.6.– Any set of n periodic synchronous tasks with implicit deadlines is guaranteed schedulable by EDF if and only if its effective processor utilization U satisfies:

$$U = \sum_{i=1}^{n} \frac{C_i}{T_i} \leq 1 \qquad [2.12]$$

Any task set that does not satisfy the above test cannot be scheduled by EDF, or any other algorithm.

In the case of constrained-deadline tasks ($D_i < T_i$), this test is merely sufficient and is based on the calculation of the effective processor load [LIU 73]:

THEOREM 2.7.– Any set of n periodic constrained-deadline tasks is guaranteed schedulable by EDF if its load factor δ satisfies:

$$\delta = \sum_{i=1}^{n} \frac{C_i}{D_i} \leq 1 \qquad [2.13]$$

In the context of EDF, schedulability analysis for periodic tasks can also be conducted using the criterion of *processor demand*. In [BAR 90], the authors derive an exact schedulability condition based on busy period analysis. They show that a set of sporadic tasks is schedulable by EDF if the corresponding synchronous system is schedulable. The interval of study is therefore the time interval $[0, H)$ for synchronous tasks and $[0, \Phi + 2H)$ for asynchronous tasks, where $\Phi = \max_{T_i}\{\phi_i\}$:

THEOREM 2.8.– A set of tasks $\mathcal{T}$ is schedulable by EDF if and only if:

1) $U \leq 1$;

2) $\forall 0 \leq t_1 < t_2 \leq \Phi + 2H : dbf(t_1, t_2) \leq t_2 - t_1$.

If we restrict schedulability analysis to periodic synchronous constrained-deadline tasks, the exact test suggested by Baruah *et al.* [BAR 90] may be written as:

THEOREM 2.9.– A set of synchronous periodic constrained-deadline tasks, such that $U < 1$, is schedulable by EDF if and only if:

$$dbf(0,t) \leq t, \qquad \forall t, 0 < t < L \qquad [2.14]$$

where $dbf(0,t) = \sum_{i=1}^{n} \lfloor \frac{t + T_i - D_i}{T_i} \rfloor C_i$

The authors showed that in the case where $U < 1$, the interval of study can be restricted to $[0, L)$, where $L = \frac{U}{1-U} \max_{i=1..n} (T_i - D_i)$. The complexity of this test is pseudo-polynomial: $O(n.\max_{i=1..n}(T_i - D_i))$.

In 1996, Ripoll *et al.* [RIP 96] established a tighter upper bound by showing that it is possible to restrict the interval of study to $[0, L^*)$, where $L^* = \frac{\sum_{i=1..n}(T_i - D_i)u_i}{1-U}$.

Still in the case where $Di \leq T_i$, Spuri [SPU 96a] and Ripoll *et al.* [RIP 96] also derived another upper bound for their interval of study, additionally proving that a deadline will necessarily be missed in this interval if the task set is not schedulable by EDF. The length of this interval L_{busy} is equal to the first synchronous busy period of the processor. It can be calculated as follows [RIP 96, SPU 96a]:

$$w^0 = \sum_{i=1}^{n} C_i \qquad [2.15]$$

$$w^{m+1} = \sum_{i=1}^{n} \lceil \frac{w^m}{T_i} \rceil \qquad [2.16]$$

The recurrence terminates when $w^{m+1} = w^m$ and $L_{busy} = w^{m+1}$.

Since there is no direct link between L^* and L_{busy}, the interval of study that must be checked can be bounded by the value $min(L^*, L_{busy})$. Furthermore, noting that the demand bound function $dbf(t)$ only changes at times with absolute deadlines, the schedulability test becomes [BAR 93, RIP 96, SPU 96a]:

THEOREM 2.10.– A set of synchronous periodic constrained-deadline tasks, such that $U < 1$, is schedulable by EDF if and only if:

$$dbf(0,t) \leq t, \qquad \forall t, 0 < t < L_{lim} \qquad [2.17]$$

où $L_{lim} = \{d_k | d_k = kT_i + D_i \wedge d_k < min(L^*, L_{busy}), k \in N\}$

In 2009, Zhang and Burns [ZHA 09] achieved a further slight improvement in the upper bound L^* and also relaxed the constraint on the relation between the period and the critical delay of tasks:

THEOREM 2.11.– A set of synchronous periodic arbitrary-deadline tasks, such that $U < 1$, is schedulable by EDF if and only if:

$$dbf(0,t) \leq t, \qquad \forall t \in P \qquad [2.18]$$

where $P = \{d_k | d_k = kT_i + D_i \wedge d_k < L^{**}), k \in N\}$ and $L^{**} = \max\{(D_1 - T_1), ..., (D_n - T_n), \frac{\sum_{i=1}^{n}(T_i - D_i)u_i}{1 - U}\}$

Although this only needs to be evaluated at times corresponding to the absolute deadline of some task in the interval $[0, P]$, the set of points that must be tested can be relatively large. Because of this, Zhang and Burns [ZHA 09] also suggested another approach in 2009 entitled *Quick convergence Processor-demand Analysis (QPA)* with the goal of reducing the set of points at which the demand bound function must be calculated. Instead of checking all absolute deadlines in increasing order, the suggested algorithm begins evaluation at the highest value of absolute deadline and "works down" to deadlines with lower values, "skipping" the evaluation of certain intermediate deadlines as specified by Algorithm 2.1.

Algorithm 2.1. QPA

$t \leftarrow d_m$
while $dbf(t) \leq t \wedge dbf(t) > min\{d\}$ **do**
  if $dbf(t) < t$ **then**
    $t \leftarrow dbf(t);$
  else
    $t \leftarrow \max\{d|d < t\};$
  end if
end while
if $dbf(t) < min\{d\}$ **then**
  the task set is schedulable;
else
  the task set is not schedulable;
end if

The EDF algorithm is a very efficient algorithm. It in fact allows an effective processor utilization of 100% to be achieved in the conditions stated in theorems 2.6 and 2.7. Note, however, that the EDF algorithm is very unstable when overloaded. Experiments conducted by Locke [LOC 86] showed that EDF experiences a *domino effect* (cascade of timing violations; tasks that miss their deadlines delay other tasks that in turn miss their own deadlines) and that its performance rapidly deteriorates on overloaded intervals. This is a result of the fact that the EDF algorithm always awards the highest priority to the task with the closest deadline. In overloaded scenarios, EDF does not provide any guarantees as to which tasks will meet their timing constraints: this behavior, which is described as indeterministic, is particularly undesirable if overloading can occur unexpectedly and intermittently, for example as a result of changes in the environment. The EDF domino effect is illustrated in Figure 2.8.

The unexpected occurrence of the task τ_1 delays the execution of the task τ_2 which can no longer meet its deadline. This delay postpones the execution of τ_3 which also misses its deadline, and of τ_4, etc.

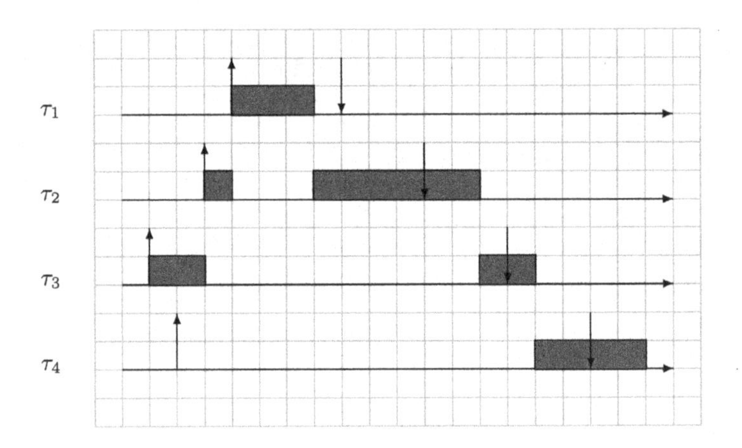

Figure 2.8. *EDF domino effect*

2.4.2.2. *Least Laxity First (LLF)*

2.4.2.2.1. Description

The Least Laxity First (LLF) algorithm prioritizes the task with the least *dynamic slack*, which is performed as follows: at each time t, the highest-priority task is the task with the smallest slack at t, defined as $L_i(t) = d_i - (t + C_i(t))$ [MOK 78].

This quantity, which is illustrated in Figure 2.9, represents the maximum time that the processor can remain idle after t without causing a missed deadline. Note that $C_i(t)$ shows the *remaining* execution time of the task at time t. $\forall t$, we must have that $L_i(t) \geq 0$.

The LLF algorithm is illustrated with a set of tasks $\mathcal{T} = \{\tau_i(C_i, T_i, D_i), i = 1..3\}$ in Figure 2.10. The tasks are assumed to be synchronous with initial activation offset $\phi_i = 0$.

This scheduling policy maximizes the minimum lateness of any task set. Unlike the EDF algorithm, LLF considers not just the urgency of the work to be performed, but also the duration of this work.

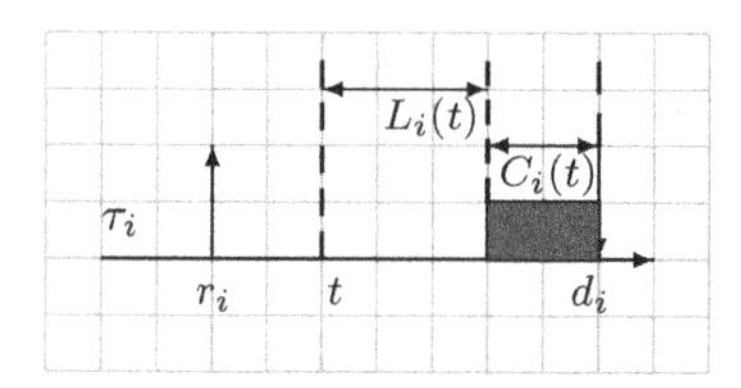

Figure 2.9. *Processor slack at time t*

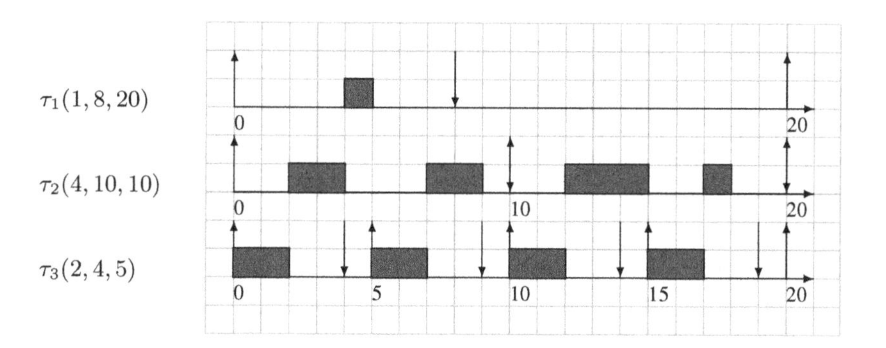

Figure 2.10. *Illustration of the working principle of LLF*

2.4.2.2.2. Optimality results

Sorenson [SOR 74] proved the optimality of this algorithm: LLF is optimal in the class of preemptive algorithms for independent periodic tasks, such that $D_i \leq T_i$.

2.4.2.2.3. Schedulability conditions

The schedulability conditions of LLF are the same as those of EDF:

THEOREM 2.12.– Any set of n periodic implicit-deadline tasks is guaranteed schedulable by LLF if and only if its effective processor utilization U satisfies:

$$U = \sum_{i=1}^{n} \frac{C_i}{T_i} \leq 1 \quad [2.19]$$

The LLF algorithm is as efficient as EDF, except that LLF has the disadvantage of potentially generating a large number of preemptions, and ceases to be optimal if preemption is disallowed [GEO 96].

2.5. Aperiodic task servers

Scheduling algorithms implemented in order to *conjointly* execute periodic and aperiodic tasks must not only be capable of guaranteeing the deadlines of critical periodic tasks, but must also provide acceptable average response times for non-critical aperiodic tasks, even though the occurrence of aperiodic requests is non-deterministic. For critical aperiodic tasks, the objective is to maximize the average acceptance rate of these requests in the system.

Two approaches are commonly encountered: *fixed*-priority task servers and *dynamic*-priority task servers, besides the traditional approach of serving aperiodic tasks as background tasks. A description of each of these approaches is given in the following sections.

2.5.1. *Approach based on background processing*

Background (*BG*) aperiodic task servers [LEH 87] schedule non-critical aperiodic tasks during times with no periodic activity (i.e. when no periodic tasks are in the ready state) within the system. The main advantage of this kind of server lies in the simplicity of its implementation, as well as the fact that it guarantees that aperiodic tasks will not disrupt the behavior of periodic tasks. However, the major disadvantage is that the response time to aperiodic requests can be prohibitive. Indeed, if the processor utilization allocated to periodic tasks is high, there will be limited opportunity for the execution of aperiodic tasks.

2.5.2. Approaches based on fixed-priority servers

The *Polling Server (PS)* [LEH 87], *Deferrable Server (DS)* [STR 95] and *Sporadic Server (SS)* [SPR 89] are the best-known examples in this category of server. They all rely on the principle of a *server task* responsible for executing aperiodic tasks. Aperiodic tasks are served when this server task is activated, according to the technique of FIFO. In the case of the Polling Server and the Deferrable Server, the server task is periodic with period T_s and execution time C_s (also referred to as the *server capacity*). For the Polling Server, if there is no pending aperiodic task when the server is activated, its capacity is wasted. The Deferrable Server behaves identically to the Polling Server, except that the Deferrable Server *conserves* its current capacity until the end of the activation period. The main problem with the Deferrable Server is that conserving the server capacity can compromise the execution of periodic tasks. The Sporadic Server solves this problem by not refreshing the capacity periodically, but instead only once it has been fully expended by aperiodic tasks.

In terms of the response time to aperiodic requests, the performance achieved is better than that achieved by handling aperiodic tasks as background tasks. Additionally, this approach has low computational complexity and is easy to implement. However, one of the disadvantages of this method lies in the sensitivity of the process of defining the server task parameters to ensure acceptable service of aperiodic tasks. Ideally, its execution time should be as high as possible, and its period should be as low as possible.

Alternative methods based on similar principles are given by the *Priority Exchange Server* [LEH 87] and the *Sporadic Server* [SPR 89] respectively established by Lehoczsky *et al.*, and Sprunt *et al.* In 1992, Lehoczky and Ramos-Thuel also suggested a method for serving aperiodic requests called the *Slack Stealer* [LEH 92]. This method is based on the idea of recovering as much time as possible from periodic tasks, without compromising their deadlines. This approach, based on the calculation of the slack of periodic tasks, was later extended by Davis *et al.* [DAV 93] and Tia *et al.* [TIA 95].

2.5.3. Approaches based on dynamic-priority servers

Most of the methods studied in connection with fixed-priority algorithms have been extended to dynamic-priority algorithms. For example, multiple mechanisms have been suggested for EDF by Ghazalie and Baker [GHA 95] (*Deadline Deferrable Server*, *Deadline Sporadic Server* and *Deadline Exchange Server*) and also by Spuri and Buttazzo [SPU 94, SPU 96b] (*Dynamic Sporadic Server*, *Dynamic Priority Exchange Server* and *Improved Priority Exchange Server*).

Another example is Earliest Deadline as Late as possible (EDL) [CHE 89], which schedules aperiodic requests at the earliest opportunity, postponing the execution of periodic tasks as much as possible, similar to the effect of the *Slack Stealer* server suggested by Lehoczky and Ramos-Thuel [LEH 92] for fixed-priority systems. To be more precise, the EDL server executes periodic tasks as early as possible in the absence of aperiodic activity. But whenever an aperiodic request arises, all periodic tasks are scheduled as late as possible, while still ensuring that all task deadlines are met. In other words, the EDL algorithm exploits the effective slack (i.e. the interval between the execution start and finish times) of periodic tasks in order to minimize the response time of aperiodic tasks. It has been shown [SIL 99] that the EDL server only needs to calculate the processor slack online at times corresponding to the occurrence of a new aperiodic task. A fundamental property of the EDL server is that it guarantees maximum slack for any set of tasks. EDL was proved to be optimal [SIL 99]. Another significant result lies in the complexity of establishing the EDL schedule. The online calculation of slack times [SIL 99] is performed in $O(\lfloor \frac{R}{p} \rfloor n)$, where n is the number of periodic tasks, R is the latest deadline of the active tasks and p is the smallest period.

The *Total Bandwidth Server* (*TBS*) [SPU 94, SPU 96b] also uses a simple mechanism to efficiently serve aperiodic tasks. Each time that an aperiodic request is registered in the system, the TBS assigns a *virtual* deadline as a function of its bandwidth (in terms of CPU execution time). Once this *virtual* deadline has been assigned to the aperiodic request, it is scheduled together with the periodic tasks by EDF. The advantage of this approach lies in the fact that the implementation overhead is essentially negligible. The optimal version of the TBS, denoted as TB* (*Optimal Total Bandwidth server*) [BUT 99], reduces the virtual deadline calculated by the TBS as much as possible in order to improve the response time of aperiodic requests while ensuring that the set of periodic tasks remains schedulable. If the average case execution time (*ACET*) of each task is equal to its worst-case execution time (*WCET*), then the TB^* algorithm is optimal in the sense that it minimizes the response time of aperiodic tasks. However, this optimality requires two other assumptions, namely that aperiodic requests are served according to the policy of FIFO, and that deadline conflicts are resolved in favor of aperiodic tasks.

2.6. Conclusion

In this chapter, we presented an introduction to the theory of scheduling in real-time systems. The goal of this chapter was first of all to provide a description of the generic model of real-time tasks on which this theory is based. We introduced the terminology and concepts associated with the topic of schedulability analysis, and listed the classification and properties of scheduling algorithms. We then presented the main algorithms for scheduling periodic tasks. In particular, we recalled schedulability conditions and optimality results for each of them. Finally, we gave a brief overview of the main servers for aperiodic tasks. Indeed, even if the regular processing load of the system is comprised of periodic tasks, aperiodic tasks can also be present.

We will see in Chapter 4 that classical real-time schedulers have been found to be unsuitable for real-time systems powered by renewable energy sources, referred to here as autonomous real-time systems. The models and concepts presented in this chapter will be extended to consider an additional dimension: energy. But first, the next chapter will give an introduction of this new generation of real-time autonomous systems arising from the proliferation of connected devices.

3

Harnessing Ambient Energy for Embedded Systems

Over the last decade, researchers and companies have been actively working to develop new methods and techniques for extracting, storing, and efficiently and economically exploiting the energy that surrounds us, often described as green energy. The process of gathering this energy is more commonly known as *energy harvesting* or *energy scavenging*. There are a number of physical phenomena that allow us to gather environmental (or ambient) energy and convert it into electrical form. The best-known methods are piezoelectricity and photoelectricity. Energy harvesting technology, still in its infancy for some types of energy, will allow wireless systems to become fully autonomous for the full duration of their lifetime [PAR 05]. In this chapter, we will begin by giving an overview of the principal forms of energy present in our surroundings, whether already exploited or on their way to becoming exploitable in the near future. We will focus, in particular, on the types of energy that are particularly suitable for powering small embedded systems, such as wireless sensor nodes and cyber-physical systems. In the next chapter, we will then be able to describe the software technology crucial for implementing autonomous real-time systems characterized by two significant global constraints: *time* and *energy*.

3.1. Why is it necessary to harvest energy from the environment?

3.1.1. *Constantly evolving technology*

Energy harvesting technology is based on the idea that devices can harvest the energy present in their ambient environment in *real time* and use it immediately, so that energy only ever needs to be stored temporarily. This could allow devices to achieve a theoretically infinite lifetime, limited only by the lifetimes of their components [CHA 08]. However, it remains to be shown that this new technology is applicable to real-time systems, as their operation is subject to the requirement that they must strictly observe specific response times.

The term *energy harvesting* is generally used in connection with supplying power to small electronic components described as *low-power* [PRI 09]. Connected objects including wireless sensors and wearable electronic equipment are the most important domains of application of energy harvesting. The implementation of these new technologies has notably prompted a shift in the design approach of electronic systems. It introduces new challenges for system designers who must now also attempt to optimize the utilization of the available ambient power to achieve energy self-sufficiency in each device. This challenge will arguably become easier to solve over time. Indeed, the power consumption of electronic circuits and wireless connections has steadily decreased. This has led to a dizzying expansion of energy harvesting technology in every field of application: home automation, medicine, military, transport, etc. By 2024, the global market of devices powered by ambient energy is expected to total 2.6 billion units. However, energy harvesting presents a new set of challenges, the majority of which can be traced back to the uncontrollable and unpredictable character of most sources of ambient energy.

3.1.2. *Terminology*

Thanks to constant advancements in micro-electronics and energy storage, we are witnessing a technological revolution driven by the key concepts of *connected embedded systems*, *smart objects*, the *Internet of Things (IoT)*, *wireless sensor networks* and *cyber-physical systems*. This revolution is based on the overarching idea of embedding a software-driven processing resource (e.g. a microcontroller) onto devices, together with the ability to communicate data about its environment, usage and state over a wireless connection. In order to clarify this general context, we will now give definitions of these concepts:

– sensor: a device that transforms a measured physical quantity (temperature, pressure, altitude) into digital form by means of a transducer so that it can be processed [MCG 13];

– connected embedded system: an autonomous electronic and data processing system, specializing in a certain specific task and equipped with a communication interface and various sensors and/or actuators. This term can refer to either the physical processing hardware, or the software that drives it. The resources of these systems are generally limited. These limitations are generally spatial (small size) and energetic (restricted power consumption). Embedded systems act as "hidden intelligence" in a wide range of products and devices (also referred to as objects), from cars to toys to robots, household appliances, machine tools and measuring equipment. They are in high demand by both the general public and the industry. We also use the term *sensor* or *sensor node* to refer to a connected embedded system without an actuator. The market value of these systems will be more than 230 billion dollars in 2020 [BAR 12];

– smart object: a physical system, generally of small size, containing a connected embedded system equipped with internal processing hardware (such as a micro-processor or a micro-controller) and a bidirectional communication interface;

– Internet of Things: an expansion of the Internet for inter-connecting objects, allowing them to interact with other objects or with us, for example over specific connections (wireless, ZigBee, Bluetooth). The Internet of Things marks the beginning of a new era in which everyday items such as toothbrushes become smart. These things will integrate transparently into our everyday lives and will be capable of generating and exchanging useful information without human intervention. Objects can also connect with smartphones, tablets and/or computers. The main advantage lies in the interactivity of all connected devices and the possibility of recovering information or transmitting statistical data [CHA 10];

– wireless sensor network: a set of multiple sensors covering a certain geographical region or a region delimited by a system of any size. Sensors can interact with each other and with an external system (e.g. the Internet) [MCG 13]. Even though they are not necessarily visible, sensors play a critical role in ensuring proper function in a large number of industrial and consumer systems. Whether ultimately used for monitoring, analyzing or controlling a process, information is originally gathered by sensors. The performance levels expected from sensors are increasing, in terms of the precision of the collected data, their miniaturization, their power consumption and their development costs. Developments in wireless sensor networks (WSNs) [AKY 02, AKY 10] for consumer or specialized markets are deeply related to the development of ambient energy harvesting solutions that are both efficient and economical at large scales;

– cyber-physical systems: connected embedded systems that exchange real-time data over the Internet [GUN 14].

3.1.3. *The ecological footprint*

The revolution of omnipresent computing raises a new set of environmental issues due to radio emissions and also as a result of the increasing energy consumption. Today, the primary concerns of system designers are producing enough energy to power wireless devices and storing this energy. Storage technology has strongly improved in the last few years. New generations of battery use lithium, which is less harmful than lead, but does not eliminate the ecological footprint. Micro-batteries have also arrived on a wide variety of markets, such as medical assistance, RFID badges and autonomous sensors. These batteries are extremely thin (less than 1 mm thick), increasingly reliable, less dangerous and more environmentally friendly as they do not contain either liquid or mercury. Still, these developments are unfolding at a slower pace than the advancements in digital memory storage, processor and wireless communication technologies. An estimated 30 billion batteries are sold yearly. The question of recycling these batteries after usage is a major issue, due to the quantity of waste that they generate. This waste can pose a serious hazard to health and to the environment.

3.1.4. *Miniaturization*

Miniaturization is also a key problem in industrial circles. The size of autonomous systems in terms of their weight and volume is a determining factor in their design. Embedding a miniaturized battery or a simple capacitor, or even no energy storage device at all, would allow undeniable savings. The increasing number of mobile devices that must be as small and as light as possible means that the size of components including batteries and sensors is a major factor. Accordingly, there are low-power digital gyroscopic sensors with a size of approximately 10 mm^3 and a power consumption of only a few milliamperes.

3.1.5. *Battery life*

Another shortcoming of current commercially available connected devices, especially wearable devices, is the limited battery life. Users value the fundamental qualities of lightweightedness and compactness in smart accessories. These qualities are only achieved at the expense of battery life, due to the restrictions placed on the size of their energy storage units. Even though processors are becoming more and more efficient, screen resolution is constantly increasing, as is the energy consumption of these screens. Future rechargeable battery technology will allow connected devices to push back current limitations: flexible batteries only 0.3 mm thick are arriving on the market, opening up new avenues of possibility for connected devices, such as smart jewelry.

There are a wide variety of options currently available to embedded systems designers. Embedded systems are typically powered by a non-rechargeable battery known as the *primary battery*. Primary batteries currently have a high energy density, of the order of

250 Wh/kg for alkaline and lithium batteries, and low self-discharge rates of less than 1% per year. Unfortunately, the stored energy is essentially proportional to the quantity of available matter. Furthermore, once the battery has fully discharged, the sensor ceases to function. The necessity of replacing discharged batteries therefore negates most of the advantages of this type of energy storage. The alternative is to use rechargeable *secondary batteries*. These batteries have lower energy densities, of the order of 100 Wh/kg for lithium-ion batteries. Other existing technologies also provide an alternative to lithium-ion batteries. Sodium-ion batteries represent one such example, available in a standard industrial form (cylinder with a diameter of 1.8 cm and a height of 6.5 cm), and are used in many embedded systems. The advantage of this approach, according to its proponents, is that the element of sodium is much more abundant than lithium and also less expensive. Its lifetime, measured as the maximum number of charging cycles before a significant loss in performance, is estimated to exceed 2,000 cycles. In the vast majority of applications of the IoT, recharging or replacing a defective battery is difficult or technologically and economically impossible. This is the most important motivating factor for developing systems that do not require any form of human intervention. The goal is to define systems that draw energy from their surroundings. Some of them, those that require very low power, will not need a battery, as they require less energy than that provided by ambient sources at any given moment in time.

3.1.6. *Energy saving*

A number of techniques have been suggested for slowing the rate of battery depletion and thus increasing the battery life of autonomous electronic systems. These techniques simply aim to reduce the power consumption. In low-power or ultra-low-power systems, the power consumption is massively reduced using innovative techniques relating to integrated circuits. *DVFS (Dynamic Voltage Frequency Scaling)* methods [LE 10] slow the processor speed, and *DPM (Dynamic Power Management)* methods [SIN 01] power off all or part of the electronic circuits. Despite these technological advancements, the maximum quantity of energy that can be stored in a battery or made available between consecutive charges remains limited. It is certainly not sufficient to guarantee the operation of an application over multiple decades, or even much shorter periods in energy-intensive cases. Apart from low-power or ultra-low-power technology, any significant improvements in longevity will be the result of ambient energy harvesting techniques.

3.1.7. *Summary*

We can summarize the most important properties of a connected system, and, in particular, those of a wireless sensor, as follows:

– lifetime: long or very long, of the order of multiple decades;

– network scalability: most sensor networks contain dozens or even hundreds or thousands of nodes;

– sensor localization: sensors can be hidden inside a component, inside the human body or in a hostile environment such as underwater. Access is almost impossible, or at the very least extremely costly and dangerous;

– cost: sensor networks can contain large numbers of nodes. The unit cost per node must therefore be as low as possible.

The enthusiasm for ambient energy harvesting is fueled by the following benefits:

– autonomy and cost (no intervention or maintenance);

– security and availability (service continuity);

– hostile or inaccessible environments (self-sustainability);

– clean energy (no toxic waste emitted into the environment);

– energy conservation (ambient energy is free and available in unlimited supply).

Autonomous system designers therefore require engineering methodology that allows them to balance the potential energy sources against the system energy requirements. Unlike software development, which can rely on access to off-the-shelf software components, no off-the-shelf energy harvesting unit can automatically provide a satisfactory solution. The classical diagram of an energetically self-sufficient device is shown in Figure 3.1. In addition to the energy harvesting unit, which collects energy produced by the surroundings, a buffer known as a *storage unit* (capacitor and/or battery) allows the electrical energy obtained by converting primary energy to be stored.

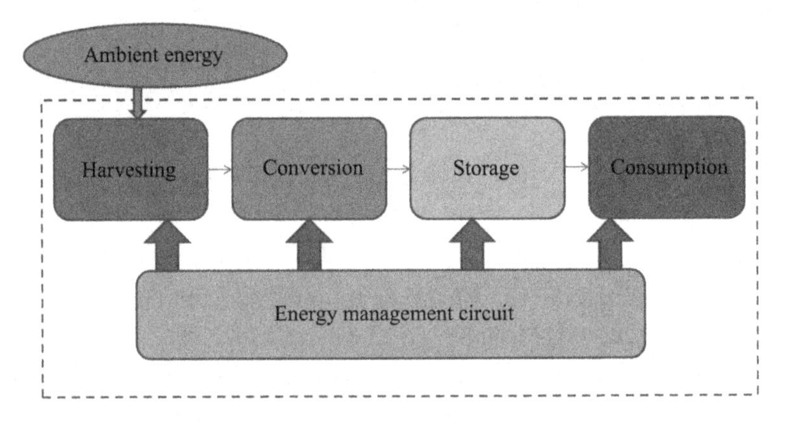

Figure 3.1. *Energy management in an autonomous system*

3.2. A wide range of applications

With a projected 50 billion connected objects in 2020, we are witnessing a complete revolution in all sectors of application: health, industry, infrastructure, transport, home, etc. In 2015, 23 million wirelessly rechargeable wearable electronic devices were sold, and 40% of all wearables sold between now and 2020 will offer wireless charging as a feature. Watches are a significant part of this phenomenon, as it is estimated that 40% of these wirelessly rechargeable objects will be watches.

3.2.1. *Infrastructure*

It is in the domain of infrastructure that connected objects will undoubtedly reveal their full potential. At a technical level, smart infrastructure combining sensors, network connectivity and software for monitoring and analyzing complex systems will be capable of detecting inefficiencies, and will provide useful information for operational decision-making. Sensors can periodically collect operational information and forward it to wherever it is required.

Connected freight carriages or containers can provide more useful information within the context of a supply chain. It is easy to imagine high-tech objects that can directly forward diagnostic information to after-sales services in order to assist in repairs and remote support. Another recent example is given by smart electricity meters, which are currently being deployed in France. These meters will automatically transmit the information required to determine consumption and billing. Finally, consider the example of smart highways. Connected sensors are embedded in the road to measure its surface temperature. External sensors measure the temperature and humidity of the surrounding air, and transmit this information to a gateway node that forwards the information over the Internet. The real-time measurements thus collected effectively complement weather projections and can be used in order to plan patrols and salting operations.

3.2.2. *Sport*

A connected sports watch or bracelet allows users to track the number of kilometers walked or ran, and provides access to statistical data, personal history and individual records. These results can be synchronized with a smartphone or tablet, or linked with online coaching sessions. This kind of connected device also provides social sharing functions, allowing users to compare and motivate themselves within a group. Some GPS-equipped watches are specifically intended for running enthusiasts. There are also sensors for golf and tennis, again with the objective of allowing users to record, analyze and improve their performance.

3.2.3. *Recreation*

Connected watches allow users to receive emails, text messages, information "in real time", access their music collection or view their photos and videos, calculate their itinerary, etc. And of course, connected televisions give access to multimedia content, viewer statistics, etc.

3.2.4. *Health*

With an area of 2.5 cm^2 and thickness of 50 microns (the same as a strand of hair), this patch contains both integrated sensors for measuring ultraviolet radiation exposure and an NFC chip for sending data to a smartphone or a tablet. The sensors are made of photosensitive colorants that act as dosimeters, changing color as a function of UV dosage. The user simply takes a photo of the patch to allow the smartphone application to analyze the collected data. Users are notified in the event of overexposure, providing *ad hoc* recommendations for sun protection. The patch is waterproof, and can be worn for several days on the back of the hand.

3.2.5. *Home automation and security*

Home automation is the sector in which we are already most familiar with concrete applications of this technology. For example, security cameras allow users to monitor their homes remotely, and generate an alert in the event of an intrusion. Connected devices make it possible to determine, adjust and optimize energy consumption. For example, a connected thermostat allows the ambient temperature to be adjusted remotely, and allows heating to be optimized around the time of day, usage hours, etc. There has also been a proliferation of integrated solutions based on sensors wirelessly connected to a hub to reduce electricity usage. These systems include, for example, presence detectors that control lighting or turn off the heating in unoccupied rooms, devices that deactivate appliances such as hot-water tanks during periods of peak consumption, and that monitor usage in real time. Thus, the fields of application are numerous and varied. But they all have one aspect in common: the necessity of measuring, processing and transmitting data in real time.

3.2.6. *Transportation vehicles*

The standard example is a pressure sensor mounted in the tires. The energy required to power the sensor is drawn from an electromagnetic signal that simultaneously relays a stream of information from the sensor. This example clearly shows the benefit of energy harvesting technology for the development of mechatronic systems in cars.

3.3. Useful energy sources

3.3.1. *Diversity and limitlessness*

Many different energy sources are currently available. Each ambient source has its own specific characteristics regarding the regularity with which it produces energy. One single source might not suffice to guarantee continuous and regular power and eliminate the risk of an energy shortage. In order to function correctly, an autonomous system must therefore exploit a combination of multiple energy sources, and must also dynamically adapt its power consumption to match the output of this (or these) source(s).

In order to compare different sources of renewable energy, the quantity known as the *volume (or surface) power density* is often used. This is the amount of power harvested per unit volume, which is measured in W/m^3. Figure 3.2 clearly shows the superiority of photoelectricity.

Figure 3.2 shows a non-exhaustive overview of the energy resources available in the environment, classified according to their characteristics.

3.3.2. *Mechanical sources*

Machine vibrations and the stress exerted on some materials are a possible source of mechanical energy [DES 13]. Mechanical energy is frequently used in medical applications, as it can be generated by the motion of the human body (walking, running, etc.). The two principal types of electrical generators are based on piezoelectrical conversion and electromagnetic conversion. Today, these techniques have been successfully implemented in embedded systems with power yields of up to several dozens of watts.

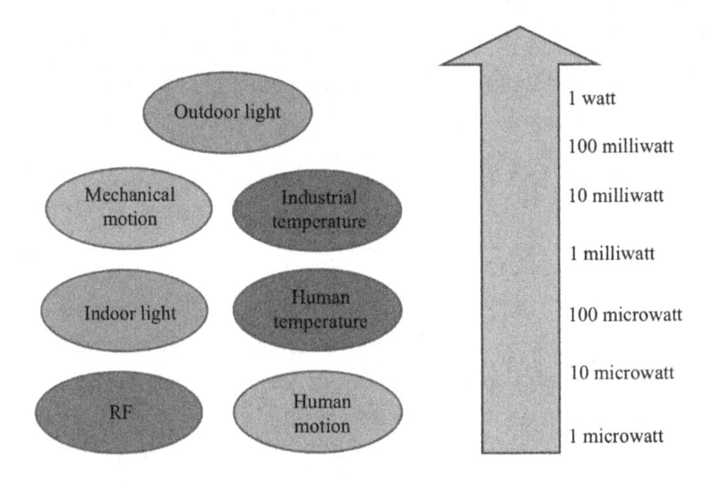

Figure 3.2. *Power densities of energy sources*

Piezoelectricity is based on the principle that certain objects can be electrically polarized by the action of a mechanical force, and conversely experience a deformation when an electrical field is applied [SOD 05]. This polarization is proportional to the stress placed on the object. There are many types of piezoelectrical materials, including simple crystals such as quartz, ceramics such as lead zirconate titanate (PZT) and polymer materials. Unfortunately, the piezoelectrical properties of materials deteriorate with the age of materials, the number of stimuli and the temperature.

Footwear initially developed by MIT recovers electrical energy from walking using a piezoelectrical generator. Another example is nanoribbons printed on silicone sheets; 80% of the mechanical energy applied to this piezoelectric material is converted into electrical energy. The biocompatibility of silicone allows this kind of sensor to be implanted inside the body in order to collect energy from a person's movements or even breathing. This technology could be used to power implanted medical devices such as pacemakers or insulin pumps.

Electromagnetic induction, discovered by Faraday in 1831, involves the generation of an electrical current in a conductor placed within a magnetic field. In most cases, the conductor takes the form of a coil. Electricity is generated by the motion of a magnet inside the coil from the variations in the flux of the magnetic field. This kind of electricity generator was, for example, integrated into a knee brace, allowing kinetic energy to be recovered from the act of walking. A person walking at a speed of 1.5 meters per second can generate approximately 5 watts of power. Another example is a wireless RF switch based on the conversion of electromagnetic energy which, once activated, uses this energy to transmit contact information, or in other words a command, to a wireless receiver by RF signal using a protocol such as ZigBee or Bluetooth low energy.

Products combining the individual benefits of piezoelectricity, which is independent of the speed at which motion occurs, and magnetism, which does not require contact and thus can go through barriers, are now seeing the light of day. Micro-generators work as follows: motion activates magnets that induce a variation in the magnetic field around a magnetic material that is deformed by the action of this field, in turn causing a deformation of a piezoelectric material, which then generates electricity. This technique is mostly used in industrial sensors, technical building management, smart gas sensors, etc.

Natural energy extracted from the motion of waves or wind has long been used to generate electricity by electromagnetic conversion on large scales. This has inspired the design of micro wind turbines to power small electronic components such as telephones.

3.3.3. *Heat*

Thermoelectricity was discovered in 1821. Thermoelectrical generators use a natural phenomenon (called the *Seebeck effect* after the German physicist Thomas Johann Seebeck) [CAR 13]. Thermogenerators contain an arrangement of junctions between two conductors held at different temperatures. A current proportional to the difference in temperature is generated in these conductors.

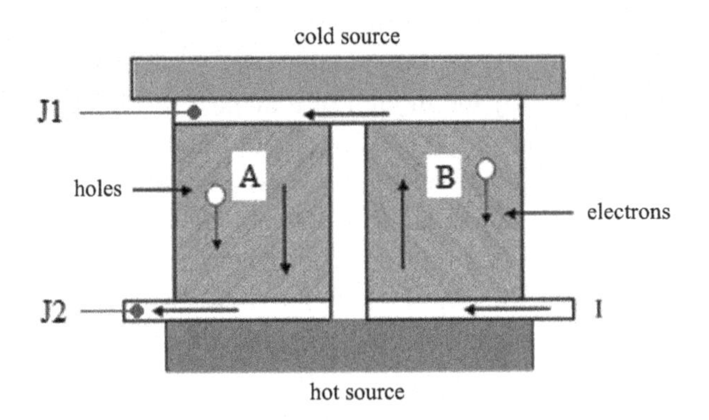

Figure 3.3. *The principle of thermoelectricity*

The number of junctions determines the quantity of electrical energy produced per unit surface. Given an effective gradient of 10 degrees, a thermoelectrical generator can produce a voltage of approximately 1 volt. Manufacturers of thin-layer components with thermoelectrical properties offer very small generators. For example, integrated into the head of a steel screw, one component can generate 15 mW with a temperature difference of 10–20°C between the material and the surrounding air.

3.3.4. *Light*

This energy source is often divided into two categories: *indoor room light* and *outdoor sunlight*. Solar energy is collected by components such as photo-sensors, photo-diodes or photovoltaic (PV) cells [KIR 15]. Solar radiation is the primary source of the renewable energy available on the planet, other examples of which include wind energy, tidal energy, bio-energy, etc. When a photovoltaic cell is oriented towards the sun, an electrical voltage is generated. Despite a significant loss during conversion, the process of extracting energy from light is accessible and reasonably inexpensive. A number of companies have designed devices that allow solar energy to be used to recharge mobile devices. Initially, this took the form of mini-solar panels that could be used to recharge devices over an integrated USB port. More recently, systems indistinguishable from the devices themselves have been designed, such as transparent photovoltaic films integrated into the terminal screen. In California, researchers are working to perfect flexible solar cells containing only 2% silicon. These cells can recover 85% of the incident solar energy, with a conversion rate of about 95%. The disadvantage, however, still remains that the energy production is irregular and depends on the season.

There is considerably less solar energy available indoors compared with outdoor environments (see Table 3.1). A model developed by Randall [RAN 04] can be used to estimate the amount of energy available indoors, taking into account all the present light sources.

Environment	Indoors				Outdoors	
	Soft lighting	Corridor lighting	Workspace lighting	Incandescent lighting	Cloudy weather	Sunny weather
E (lux)	80	150	740	1200	3030	100000
I ($\mu W/cm^2$)	67	125	616.7	1000	2525	100000

Table 3.1. *Indoor photovoltaic energy [WAL 11]*

Consider a north-facing sensor powered by solar energy, in winter, with only 6 h of daily exposure. These conditions correspond to a power density of 20 W/m^2. A solar cell of 1 cm^2 with a 12% yield can collect 5.2 J/day in these conditions. Given that current autonomous systems require about 4 J to measure and then transmit data, we can guarantee that this autonomous sensor will be energetically self-sufficient in all weather conditions [WAL 11].

3.3.5. *The human body*

Whether at rest or in motion, the human body produces energy, which is measured in calories per hour. [STA 96] gives a list of the power consumed by various activities such as sleeping, standing, playing the piano, swimming, etc. For example, driving requires about 160 W and swimming requires 580 W. Human energy production is associated with loss in the form of emitted heat. This yield, evaluated at 25%, allows us to estimate the quantity of energy that could effectively be recovered for each type of activity. Thus, 40 W could be recovered while driving and 150 W while swimming.

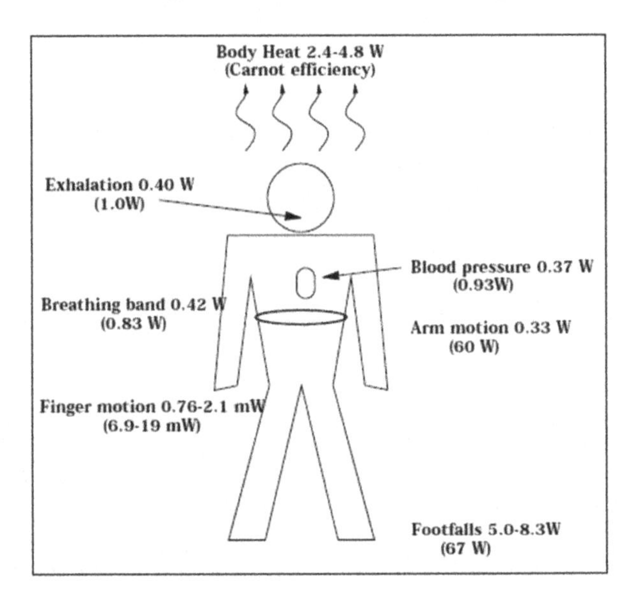

Figure 3.4. *Energy dissipated by the human body [STA 96]*

The human body regulates its internal temperature to about 37°C, dissipating vast amounts of energy (see Figure 3.4). Thus, if the temperature of the surrounding air is lower, for example 20°C, a heat flux of the order of 8 W/m^2 flows from hotter regions to cooler regions. Given a temperature difference of 17°C between the body and the surrounding air, we can conclude that bare skin dissipates approximately 140 W/m^2, or 14 mW/cm^2.

One first example of an application is a watch with an embedded thermoelectrical generator inside. The generator harnesses the temperature difference between the body and the surrounding air to power the micromotor in the watch.

A device based on piezoelectrical ceramics placed on the joints and built into a prosthesis can already generate several mW of power.

Research has also examined the question of harvesting biomechanical energy within the body itself, providing an incredible way of powering medical implants such as pacemakers and drug pumps. Not only does this technology allow these devices to operate fully independently, but also it removes the need for high-risk surgery to replace the batteries. In this way, it is possible to harvest energy from the motion of the heart, the diaphragm or the lungs by piezoelectrical conversion to power a pacemaker.

3.3.6. *Electromagnetic radiation*

Ambient electromagnetic waves emitted by power cables, mobile phones, Bluetooth connections, microwave ovens, etc. can be used to generate electricity [PEN 11]. Currently, the electromagnetic waves saturating our environment increase by 15% each year, due to widespread utilization for data transmission. This increase provides a guaranteed source of energy that is both sufficient and stable, and which may be harnessed by means of a special antenna and converted into continuous current. Both radar and FM frequencies can be used, over a spectrum ranging from 100 Mhz to 15 Ghz. Thus, it is possible to generate hundreds of μW of electricity by using television frequencies, which would be sufficient to power small electronic devices such as sensors. In 2013, Duke University in the United States presented a method of converting WiFi signals into electricity using metamaterials with a similar yield to solar cells, allowing a mobile phone to be charged with the signal emitted by a relay.

3.4. Energy storage

3.4.1. *Introduction*

It is important to note that ambient energy is often random and intermittent. In most cases, it is therefore necessary to include a way of storing electrical energy after it is generated, which must also manage the various energy levels required by each electronic component in order to function properly. The component that allows electrical energy to be temporarily stored is called an *energy storage unit*. Electricity is a secondary form of energy in the sense that once it has been created, it is more difficult to store than other forms of energy. Generated electricity is either immediately consumed, lost or stored in a capacitor. In order to utilize energy resources as fully as possible, and minimize losses, it is essential to store the energy harvested from the environment. Storing energy is a fundamentally challenging technical problem, as it often involves complicated processes that also require energy. Regardless of the technology employed [LIU 12], energy storage units are essentially characterized by three values:

– the weight (or volume) energy density, measured in watt-hours per kilogram (or watt-hours per liter). This is the quantity of energy that can be stored per unit weight (or volume) of the accumulator;

– the weight power density, measured in watts per kilogram. This is the power (electrical energy produced per unit time) yield per unit weight of the accumulator;

– the deep recharge cycle number, simply measured as a number of cycles, characterizing the lifetime of the storage unit. This is the number of times that the unit can output an amount of energy higher than 80% of its nominal energy capacity.

The operating cycle of an energy storage unit is comprised of two processes: loading and unloading. In each cycle, energy is converted twice. The yield of the storage unit is therefore particularly important, and strongly depends on the type of storage and physical hardware. With the discovery of new materials and advancements in manufacturing technology, new energy storage processes are constantly being developed. In the following, we will describe the three categories of an energy storage unit.

3.4.2. *Rechargeable chemical batteries*

Rechargeable batteries are described as secondary to distinguish them from non-rechargeable, single-usage batteries. Here, we will focus on secondary battery technologies. A battery, or more precisely a battery cell, is comprised of three parts: a positive electrode, a negative electrode and an electrolyte. These three parts are "sandwiched" to form an electrochemical system capable of producing an electrical current. Regardless of the number of cells, batteries have a limited lifetime. This is primarily due to the usual deterioration of the materials over time and of the energy that they store and redistribute.

The constituent materials of an electrode are usually compact powders or solids. The electrolyte is often a liquid, which requires containment within a porous film or as a gel to avoid spilling. The chemical composition of batteries cells varies widely, but all batteries have two fundamental characteristics, which are specified by the manufacturer:

– the nominal voltage, which determines the available electrical power, measured in volts;

– the nominal capacity, which determines the current that a battery can produce over time, measured in ampere-hours (Ah).

Having an understanding of the various different battery technologies is useful for autonomous system designers, as the choice of chemical component determines the performance and limitations of the battery. These technologies can be summarized as follows:

– nickel-cadmium batteries (NiCd) have previously dominated the market of batteries for portable electronic devices, offering mature technology at a low price and high-performance level. However, these batteries experience a self-discharge rate of about 20% per month at 21°C;

– nickel-metal hydride (NiMH) technology has begun to replace NiCd in many consumer products. These batteries render products more economically competitive, with performance exceeding that of NiCd batteries. They can achieve energy densities twice as high as NiCd batteries;

– lithium-ion batteries (Li-ion) have even higher performance still. Their cost is decreasing, thanks to their success in portable computer devices.

Chemical battery technology has improved significantly. However, these batteries require bulky packaging, which reduces their net energy density. The use of highly flammable liquid electrolytes based on organic solvents contributes to a decrease in performance over time. These electrolytes also deteriorate over time and as a function of temperature (in addition to the risk of explosion). The risk of electrolyte leakage renders these batteries ineligible for medical devices implanted inside the human body. Still, of all the technologies for storing electrical energy, Li-ion batteries continue to prove popular in a large number of applications, as they provide the best overall compromise, especially in terms of energy density and power.

3.4.3. *Supercapacitors*

Traditional batteries are characterized by high energy density and low power density. Their lifetime in terms of the number of charge/discharge cycles is relatively low. Technological advancements and improvements in materials manufacturing have led to new kinds of electrical energy storage such as supercapacitors [VAN 13]. Supercapacitors can be used as a complement to batteries or fuel cells. By means of an arrangement that is complementary in terms of the instantaneously available power and the quantity of stored energy, the performance of power supply units can be increased. Supercapacitors are also examples of electrochemical capacitors, except that they are reinforced by a dual-layer architecture (see Figure 3.5).

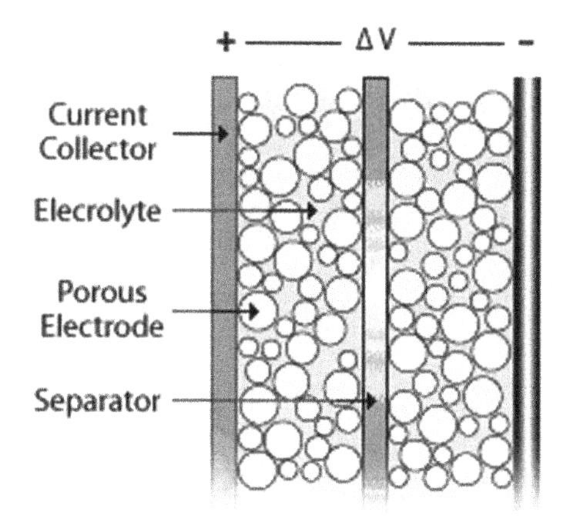

Figure 3.5. *Diagram of a supercapacitor*

Compared with rechargeable batteries, supercapacitors have the following properties:

– fast charge/discharge;

– they can produce regular pulses of energy without the risk of damage;

– extended lifetime: supercapacitors can charge/discharge thousands of times;

– lower internal resistance: efficiency levels can achieve 84–95%;

– they can be charged to any percentage, without incurring a memory effect;

– large range of operational temperatures.

However, supercapacitors are not perfect due to their high self-discharge rates and the fact that the voltage output varies as a function of the quantity of stored energy. Additionally, supercapacitors require more space than a battery of equal storage capacity. The chart in Figure 3.6 provides a brief comparison of the relative performance of the various techniques of energy storage.

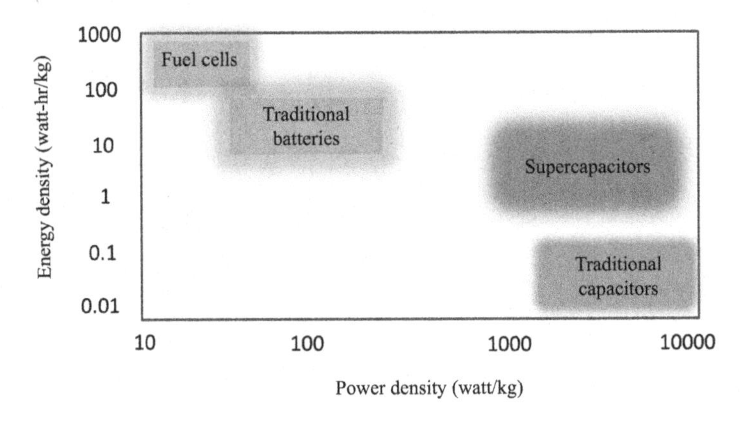

Figure 3.6. *Ragone chart*

3.4.4. *Solid-state rechargeable batteries*

Solid-state batteries offer similar benefits to those that led to the success of Li-ion batteries, with increased energy and power density, longer lifetimes, zero safety hazards and no dangerous chemicals in order to meet the constraints of ecodesign. Micro-batteries manufactured by vacuum processes derived from microelectronics are now available. With a thickness of the order of a fraction of a millimeter, they can, for example, be integrated into smart labels. They can operate for 5,000 cycles and have low self-discharge rates of less than 3% per month. Liquid electrolytes are replaced by electrodes and thin-film electrolytes. Li-polymer batteries (sometimes called Li-Po), which surfaced shortly after the year 2000, consist of an electrolyte inside a polymer material. Although they are more expensive, they are attractive as they can take on very thin and varied shapes, and placed on flexible platforms. They are lightweight, and are safer than Li-ion batteries as they are less susceptible to overcharging. They are therefore well adapted to applications of energy harvesting.

The development of these batteries creates new possibilities for miniaturization, improving the energy density, lifetime, safety and design flexibility.

In summary, compared with traditional batteries, solid-state batteries have the following qualities:

– very high energy density;

– low self-discharge rate;

– high number of charging cycles;

– extremely small size and thickness;

– no chemical or explosive hazard.

	Li-ion battery	Supercapacitor	Li-ion polymer battery
Charging cycles	Low (100)	Very high	High (>5000)
Self-discharge	Medium	High	Very low
Charging rate	Slow	Fast	Fast
Size	Medium	Large	Small
Output variation		Varied	Stable
Capacity	0.3–10 mAH	10–100 μAH	12–85 μAH

Table 3.2. *Comparison of storage technologies*

The development of connected objects that rely on energy harvesting faces a major technical obstacle: the continuous storage/output of energy by a storage unit with small dimensions, an augmented lifetime, and sufficient energy and power densities. Given what we have seen in the above, the solution to this technological rift lies in a new generation of solid-state rechargeable batteries, likely structured in three-dimensional thin layers.

3.5. Implementation of an autonomous system

3.5.1. *The components of an autonomous system*

As shown in Figure 3.7, a wireless sensor node powered by ambient energy contains the following components:

– transducer: this converts the energy received in a certain form (e.g. mechanical, thermal, light) into another form suitable for storage;

– power management unit: this oversees energy conversion, energy storage and power supply;

– sensor: this detects and quantifies ambient parameters such as temperature, pressure, pH, proximity;

– processor (or micro-controller): this receives the signals from the sensors, and converts them into useful form for analysis, processing and communication via the wireless transmitter;

– wireless radio module: this relays the information output by the processor to a data collection point either continuously, periodically or according to an event-based paradigm.

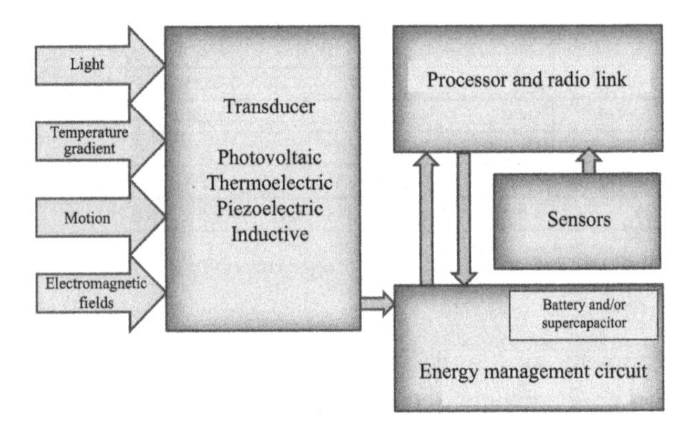

Figure 3.7. *Components of a wireless sensor node*

3.5.2. *Energy requirements analysis*

When designing an autonomous electronic system, it is crucial to analyze in advance the energy and power that it requires to function correctly [RAG 05]. This involves:

– selecting the energy source(s) and type of transducer;

– characterizing the power output of the transducer in various operating conditions;

– studying all system components to calculate the required power in all modes of operation (standby, processing, transmission);

– identifying the electronic hardware necessary for energy conversion and power management, and integrating the power consumption into the system;

– choosing an energy storage unit capable of meeting the requirements in terms of power output and storage capacity.

3.6. Current operating principles

Two categories of system currently co-exist, namely *harvest-store-use* systems and *harvest-use* systems.

3.6.1. *Harvest-store-use systems*

These systems harvest energy by one of the processes described above, and accumulate energy over an extended period of time. The energy storage unit must therefore have low rates of loss and a sufficient storage capacity. Energy-neutral operation is the result of a good balance between the average power output of the ambient source and the average power consumption of the electronic device. Maintaining a small energy reserve allows for some flexibility of operation, avoiding energy shortage over short periods of peak consumption. In general, these systems remain in a state of very low consumption known as sleep mode (or standby) for most of the time, but harvest energy continuously.

The standard example is that of a sensor node equipped with a photovoltaic cell. The node functions intermittently, alternating cyclically between standby and active modes. In the latter mode, sensor data are captured, processed by the micro-controller and then transmitted over the wireless network. The period of this cycle could be seconds, minutes or even longer. This mode of operation allows the average power consumption of the device to be reduced. Suppose that we estimate the energy required by the capture–processing–transmission chain to be 100 μJ, and the average harvested power to be 10 μW. Then, an autonomous sensor node would achieve energy-neutral operation whenever the period of the activity cycle is

10 s or longer. If we choose to implement a photovoltaic cell 10 times as large, this period can be reduced to 1 s. Essentially, this approach assumes that the waiting period between two operating sequences is large enough to allow a sufficient quantity of energy to be stored, and also small enough to meet the time constraints on the sensor activity.

3.6.2. *Harvest-use systems*

These systems remain in an unpowered sleep state until an energy impulse is detected. Upon activation, the system carries out its functions using the collected energy. Continuous operation is guaranteed by matching the periods of energy consumption to the periods of energy harvesting. The standard example is the wireless light switch, in the form of a push-button, which captures the mechanical energy generated by the applied pressure. The inductive energy converter contains an electromagnetic generator, which modulates the magnetic flux in the coils by the motion of a magnet, creating an electrical impulse that activates the wireless communication module.

3.7. Conclusion

Many devices only require a small amount of power to function. For example, medical implants can be powered with just a few dozen microwatts. Mobile devices such as phones require closer to the order of a watt, but can also be recharged by means of so-called energy harvesting systems. The energy consumption of all mobile devices is currently following a clear decreasing trend. The power consumption of the electronic circuits used in embedded systems can be expected to decrease further still over the next few years, rendering these devices more easily compatible with ambient regenerative energy. Whether mechanical, thermal, chemical, radiation-based or even nuclear, the energy available in the environment can always be converted from one form to another, and in particular into electricity.

The projected number of devices powered by renewable energy will reach 2.6 billion in 2024. Many of the applications of the IoT will involve objects installed in remote or hidden locations. Replacing the batteries of these devices is not viable, as it is too expensive or technically impossible. Thanks to high levels of efficiency in clean energy conversion (up to 94%), applications of the IoT will continue to proliferate. We are now capable of exploiting all types of regenerative energy, even in very small amounts, with minimal loss. Connected objects are characteristically subject to significant constraints on the production costs and size (weight and volume), as well as the memory and processing capacity. It is therefore crucial to precisely characterize the full chain of ambient energy production, storage and conversion into electricity in order to achieve energy-neutral operation and optimally utilize the available ambient energy [PAR 05].

The next chapter discusses embedded real-time systems powered solely by energy harvested from their surroundings, with the objective of presenting their special operational features, and accordingly demonstrating the necessity of implementing new techniques of real-time scheduling.

4

Energy Self-sufficiency and Real-time Scheduling

Designing an autonomous embedded system such as a wireless sensor node means being able to guarantee before deployment that it can achieve a mode of operation described as energy-neutral. The system must satisfy two requirements. It must never consume more energy than it gathers from the ambient source. And it must not waste energy. More precisely, the question that we will answer in this chapter may be stated as follows:

– *how can we choose the processor busy periods, and hence the idle periods during which it is placed on standby to save energy, in such a way as to prevent energy shortage?*

– *how should we schedule jobs during the busy periods so that their time constraints, expressed in terms of the latest possible time of end of execution (deadline), will always be satisfied?*

The answer to each of these equations requires finding a suitable technique for simultaneously exploiting the processing resource, which is either a processor or a microcontroller, and the energy resources present in the environment. By means of a formal study, we will show the necessity of equipping real-time operating systems with an explicit power management component responsible for dynamically switching between modes. The objective of this component is to continuously adapt the power consumption of the processing circuit to the energy production profile. The difficulty of this task lies in ensuring that the deadlines associated with each job are met despite fluctuations in the supply of available energy.

We will begin by briefly describing existing integrated methods for managing energy-neutral operations. We will then describe the underlying model of our study and the associated terminology. After demonstrating that the traditional real-time schedulers described earlier in this book are unsatisfactory, we will present a scheduling algorithm that is optimal in the context of the considered model, combining dynamic power management and real-time scheduling. We will end the chapter with a presentation of another real-time scheduler, which is also optimal, but subject to a more restrictive set of assumptions. This contrast will help to illustrate the benefit of our approach, and its wider scope of application.

4.1. Time and energy: a dual constraint

An embedded system and, in particular, an autonomous smart sensor must be capable of operating for several years or even decades without requiring maintenance. Therefore, it is crucial to establish an offline guarantee that the system will meet its constraints. The implementation of a schedulability test is complicated by the uncertainty in the quantity of harvested energy. As mentioned in the previous chapter, a large number of sensor nodes currently exhibit a very strict intermittent mode of operation, based on a prespecified *duty cycle*. The electronic system cyclically enters a phase described as sleep mode in which its consumption is minimized to a few nanoamperes as a trade-off for the fact that it cannot do any work. The sensor is programmed so that its activation occurs on a strictly periodic basis, for example once per second. The energy harvested during the sleep phase is used to power the system during the busy phase.

The drawbacks of this approach mostly lie in the difficulty of choosing the cycle when the power emitted by the ambient source varies over time, or if the tasks to be executed are not periodic. We wish to present a new approach, more flexible and less restrictive, removing the necessity of the two restrictive hypotheses of power invariance and strictly periodic processing.

The technological challenges of energy self-sufficiency were formally raised in [KAN 07] and [KAN 06]. The work of Kansal *et al.* focuses on sensor nodes powered by outdoor solar energy. A method of dynamically adapting the duty cycle of nodes is presented. This method was experimentally tested in 2005 with a platform called Heliomote. However, this work does not discuss software for real-time applications subject to strict time constraints.

Indeed, designing an autonomous real-time system raises several fundamental questions regarding the optimality and feasibility of schedules. If we assume that a perfect characterization of the power source is available (profile of the energy source, size of the storage battery, etc.), these questions may be stated as:

– *how can we check and guarantee before deployment that the system will be indefinitely autonomous, i.e. energetically self-sufficient with a consistently acceptable level of performance?* To achieve this, we must specify a *Quality of Service* parameter from the constraints of the application;

– *can the Earliest Deadline First scheduler described in Chapter 2, which was shown to be optimal in the absence of energy limitations, provide a satisfactory solution to the question of energetically self-sufficient real-time scheduling?*

– *does there exist an optimal real-time scheduler with a reasonable implementation complexity in this new context of operation based solely on ambient energy harvesting?*

In this chapter, we will consider a firm real-time system such that the quality of service is first and foremost determined by the ratio of effectively satisfied job deadlines. To answer the questions stated above, we will introduce the underlying model of our approach in the next section.

4.2. Description of an autonomous real-time system

4.2.1. *Hardware architecture*

The hardware architecture of any energetically self-sufficient embedded system such as a wireless connected device has several functional blocks, as shown in Figure 4.1. One natural approach is therefore to construct a system model based on the three following components:

– *energy harvesting unit*, the nature of which depends on the type of ambient energy, the quantity of energy required, etc.; this is the hardware responsible for extracting energy from an external source (solar, wind, vibrations, kinetic, chemical, etc.);

– *energy storage unit* such as a battery or a supercapacitor, which is chosen based on the system dynamics, size and/or cost constraints, etc.;

– *energy consumption unit*, which in this case is given by the processing and communication resources. This could, for example, be the microcontroller and the radio emitter/receiver module.

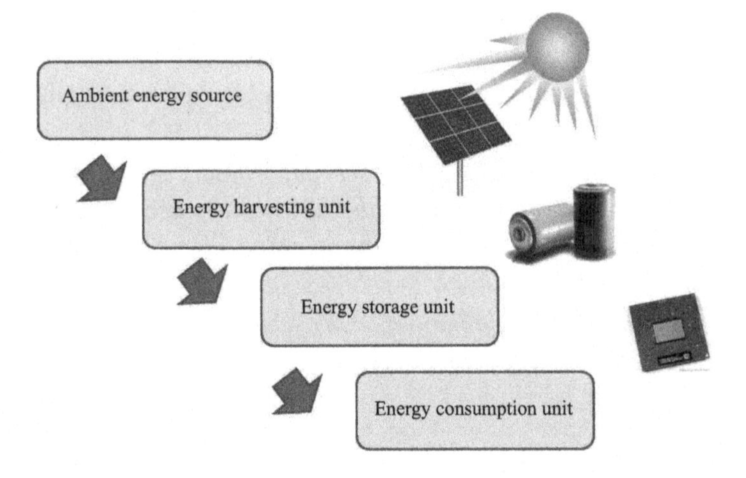

Figure 4.1. *Simplified representation of an energy harvesting device*

In this chapter, we will assume that the energy consumed by the operational part of the embedded system (actuator, sensor, LED, etc.) originates from a separate power source, as well as the radio emitter/receiver module. The energy consumption unit is therefore the electronic circuit board enveloping a microcontroller or a microprocessor. We will focus on the energy requirements of the real-time jobs executed on the processor, which consequently vary as a function of time.

4.2.2. *Software architecture*

From the software perspective, a real-time system consists of application tasks and the real-time operating system (RTOS) responsible for scheduling them. In Chapter 2, we gave an overview of the real-time schedulers typically implemented in current RTOSs. These schedulers generally have the following properties: they are online, non-idling, priority-based and preemptive. Their implementation does not present any particular difficulty: one or several data structures must simply be managed in the form of lists. The role of the scheduler is to order and update these lists, either by Rate Monotonic Scheduling with fixed priorities or with dynamic priorities using Earliest Deadline First [LIU 73]. However, these optimal schedulers operate with the assumption that energy is unlimited. Their optimality requires the processor to have access to the energy required to execute a job at any given moment. The only constraint managed by the scheduler is temporal. The schedulability conditions associated with these schedulers are therefore based on the processor utilization ratio or the processor demand over a certain interval of time.

4.2.3. *The RTEH model*

In the following, our model of an autonomous real-time system will be described as *RTEH (Real-Time Energy Harvesting)*. The RTEH model consists of a processing unit, a set of jobs, an energy storage unit, an energy harvesting unit and an energy source (see Figure 4.2).

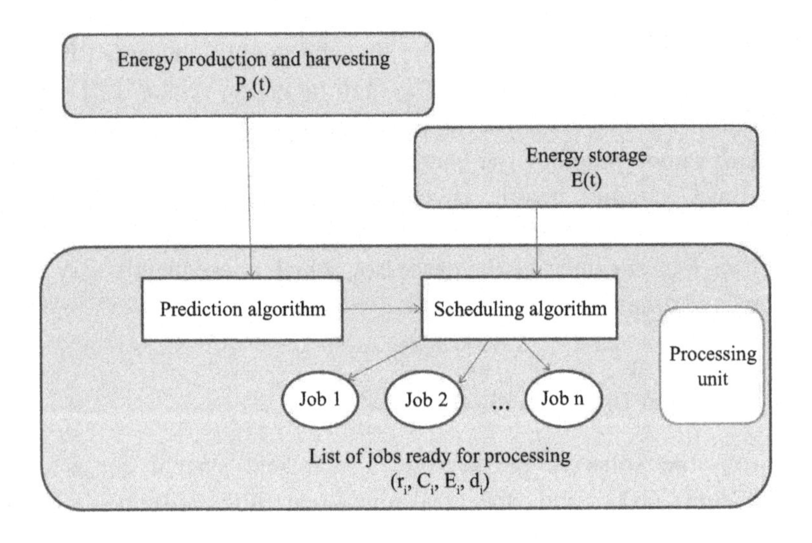

Figure 4.2. *Model of an autonomous real-time system*

4.2.3.1. *Jobs*

We will consider a set of real-time jobs executed on a uniprocessor processing unit. This unit is assumed to support one single clock rate. The energy consumed on standby is assumed to be negligible. The energy consumption of the processor originates solely from dynamic switching. Jobs are executed using only the energy generated by the ambient source. We denote $\tau = \{\tau_i, i = 1, \ldots, n\}$ the set of n preemptible jobs. These jobs are mutually independent. We assign a quadruplet (r_i, C_i, E_i, d_i) to each job τ_i. The job arrives at time r_i called activation time, has a worst-case execution time of C_i units of time and has a worst-case energy consumption of E_i units of energy. The quantity E_i is not necessarily a function of C_i [JAY 06]. In other words, the effective energy consumption of a job does not depend linearly on its effective execution time. For each unit of time, an upper bound is known for the energy consumption of any job, equal to e_{Max} units of energy. The exact quantity of effectively consumed energy in any unit of time is, however, not known beforehand. The deadline of τ_i, denoted as d_i, is the time at which τ_i must have finished execution. We assume that $\min_{0 \leq i \leq n} r_i = 0$. Let $d_{Max} = \max_{0 \leq i \leq n} d_i$ and

$D = \max_{0 \leq i \leq n} (d_i - r_i)$ respectively denote the largest absolute deadline and the largest relative deadline of the jobs in $\mathcal{T}$. $E_c(t_1, t_2)$ denotes the energy consumed by jobs over the time interval $[t_1, t_2)$. E_{tot} is the total energy required to fully execute all jobs in τ. We make the following assumptions: the energy extracted from the source in each unit of time becomes available for consumption at the end of the quantum. However, the energy consumed by a job in a unit of time must be available at the start of the quantum. The energy that is produced and consumed in each unit of time can be expressed as an integer number of energy units. If the energy consumed by a job in each unit of time is greater or equal to the energy harvested in that unit of time, we will describe the job as *discharging* [ALL 01] . All of the jobs in τ are discharging. It follows that the remaining capacity of the storage unit cannot increase while a job is executing.

4.2.3.2. *Energy production*

The energy produced by the ambient source is assumed to be uncontrollable. We will characterize it by means of the *instantaneous charge rate*, also called the production power, written as $P_p(t)$. This includes all losses incurred during the conversion and storage process. The energy produced by the source over the time interval $[t_1, t_2)$ is written $E_p(t_1, t_2)$. It can be calculated with the formula: $E_p(t_1, t_2) = \int_{t_1}^{t_2} P_p(t)dt$. We assume that energy production and consumption can occur simultaneously. The power of the ambient source can be predicted in the short term with a negligible overhead in terms of processing time and energy.

4.2.3.3. *Energy storage*

The studied system uses an ideal energy storage unit with a nominal capacity C expressed in units of energy such as joules or watt-hours. The capacity can be smaller than the total energy consumption of a job. We denote $E(t)$ the remaining capacity of the storage unit at time t, which corresponds to the current quantity of available energy. If the energy storage unit continues to be charged while full, the excess energy is considered to be wasted. Conversely, the storage unit is considered to be completely empty at time t if $0 \leq E(t) < e_{Max}$, written as $E(t) \approx 0$. The application begins with a fully charged storage unit (i.e. $E(0) = C$). Stored energy does not dissipate and may be consumed immediately after production or at any time thereafter.

4.2.4. *Terminology*

In this section, we will introduce new definitions that will be required later in the chapter. A job τ_i misses its deadline if one of the following two situations occurs:

– *time shortage*: for τ_i with a deadline at time t, execution may be incomplete if τ_i does not have sufficient time to finish executing before the deadline. However, the storage unit contains sufficient energy when the deadline is missed, so $E(t) > 0$;

– *energy shortage*: for τ_i with a deadline at time t, execution may be incomplete if there is not enough energy available to finish executing before the deadline. The storage unit is empty when the deadline is missed, so $E(t) \approx 0$.

We say that a schedule Γ of $\mathcal{T}$ is *valid* if all of the deadlines of the jobs in $\mathcal{T}$ are met when Γ begins with a full storage unit. A system for which at least one valid schedule exists is said to be *feasible*.

The limiting factor of an infeasible system could be time, energy or both time and energy. Analogously to the classical theory of real-time scheduling recalled in a previous chapter, a scheduling algorithm is said to be:

– *optimal* if it finds a valid schedule whenever at least one valid schedule exists;

– *online* if its decisions are taken dynamically;

– *semi-online* if it is online but requires future knowledge of a certain interval of time;

– *ld-omniscient* if it is semi-online and has omniscient knowledge of the ld next units of time;

– *idling* if it has the option of leaving the processor idle when at least one job is ready for processing. Otherwise, it is said to be non-idling;

– *clairvoyant* if it has full knowledge of the future (characteristics of activated jobs and energy production profile) at all times including the initial time.

We also introduce some terminology specific to the RTEH model. A schedule Γ of $\mathcal{T}$ is said to be:

– *time-valid* if all jobs of $\mathcal{T}$ meet their deadlines in Γ, with $E_i = 0$ $\forall i \in \{1, \ldots, n\}$;

– *energy-valid* if all jobs of $\mathcal{T}$ meet their deadlines in Γ, with $C_i = 0$ $\forall i \in \{1, \ldots, n\}$.

A set of jobs $\mathcal{T}$ is *time-feasible* if there exists a time-valid schedule of $\mathcal{T}$ and *energy-feasible* if there exists an energy-valid schedule of $\mathcal{T}$.

If a scheduling algorithm A requires knowledge of the energy harvested in the future in order to make a decision, we say that it is *energy-clairvoyant.*

4.3. Key theoretical results

4.3.1. *Shortcomings of* EDF

Let us consider the *Earliest Deadline First (EDF)* scheduler, which is optimal for scheduling independent jobs in the absence of any energy limitations or processing overload [LIU 73, DER 74]. EDF dynamically selects the ready job with the earliest deadline. Online schedulers such as EDF adaptively manage the variations in processor demand, and, in particular, the unpredictable arrival of jobs associated with sporadic or aperiodic tasks. The classical implementation of EDF is greedy: the processor cannot be idle if at least one job is ready for processing. In other words, the EDF scheduler can be combined with a dynamic power management policy that places the system on standby in order to save energy if no jobs are ready.

Let us briefly give a few definitions that are useful for evaluating the performance of online schedulers. The *value of a job* determines its contribution to the global system performance. The system is awarded points equal to the value of each job that finishes execution before its deadline. The system is not awarded points for jobs that miss their deadlines [BAR 91, BUT 05]. We say that an online scheduler has *competitive factor* r $(0 < r < 1)$ if it guarantees total cumulative points greater than or equal to r times the points achieved by the best possible clairvoyant scheduler [BAR 92]. An online scheduler is said to be *competitive* if it has a competitive factor strictly greater than zero. Otherwise, it is *non-competitive*.

The optimality of EDF holds so long as preemption is allowed between jobs, and these jobs are not in competition for access to shared resources. However, in the event of processing overload, EDF ceases to be optimal. Suppose that the value of each job is taken to be proportional to its execution time. Baruah *et al.* show that no online scheduler including EDF can guarantee a competitive factor greater than 0.25 in an overloaded situation for jobs with uniform value densities [BAR 92]. A recent analysis showed that in the case of the RTEH model, the competitive factor of EDF is equal to zero [CHE 14b].

THEOREM 5.1.– *EDF is a non-competitive scheduler for the RTEH model.*

4.3.2. *Illustrative example*

We will show the inadequacy of the EDF scheduler by means of a simple example. Consider a set of two jobs given by τ_1 and τ_2, such that $\tau_1 = (0, 3, 24, 8)$ and $\tau_2 = (2, 1, 6, 3)$. Energy is harvested from the surroundings with constant power over time, equal to $P_p = 2$. The energy storage unit has a capacity of $C = 8$ and is initially fully charged ($E_c(0) = 8$). Inspecting the schedule constructed by EDF, we note that τ_2 misses its deadline at time 3 (see Figure 4.3). The problem with EDF lies in its greedy energy consumption. This leads to an energy shortage for the job τ_2, which fails to finish execution before the deadline. This example shows that, despite sufficient ambient energy and sufficient processor time, the EDF scheduler does not meet the deadline constraints. The same is true for all non-idling schedulers integrated into conventional operating systems, as they are too rigid to adapt to variations in the available energy.

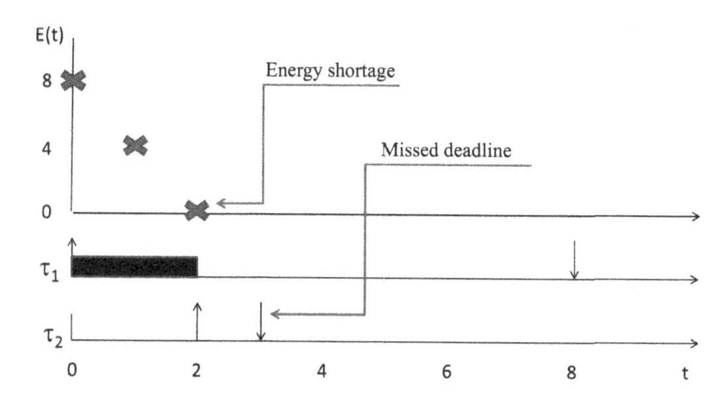

Figure 4.3. *Schedule constructed by* EDF

4.3.3. *Strengths of EDF*

The EDF scheduler is traditionally implemented in its non-idling version, called *EDS* or *ASAP (As Soon As Possible)*, in which the processor becomes busy as soon as at least one job is ready for processing. There is, however, another version called *EDL* or *ALAP (As Late As Possible)*, in which jobs are executed as late as possible without missing a deadline. In this form, the processor can be deliberately kept on standby even if jobs are ready. This version of EDF, described as idling, was previously used to optimize the management of processing time in the presence of non-critical aperiodic tasks with the goal of minimizing their response time. We will see in the next section how the ALAP mode may be used to avoid energy shortages.

EDF has a number of qualities: easy to implement in its ASAP form, fast execution of the scheduling algorithm (low overhead) and fewer preemptions compared with other schedulers. It is therefore natural to ask whether the EDF scheduler, despite being non-competitive, might nonetheless be the best non-idling scheduler for RTEH. This result is indeed established in theorem 5.2 [CHE 14b].

THEOREM 5.2.– *EDF is optimal in the class of non-idling schedulers for the RTEH model.*

We conclude that the EDF scheduler is still the best choice for implementation in systems that do not support dynamic power management, in which non-idling schedulers cannot be used. Additionally, its implementation does not require any particular technology: it does not need to know the current amount of energy in the storage unit, nor does it require a predictive estimate of the energy that will be harvested in future.

4.3.4. *Necessity of clairvoyance*

From the above, we deduce that any energy shortage (i.e. a situation in which there is insufficient energy to complete the execution of a job within its deadline) must be anticipated sufficiently early to avoid causing a missed deadline. In other words, in order to improve the performance of a scheduler, it must be made clairvoyant [CHE 14c].

THEOREM 5.3.– *In the RTEH model, any optimal online scheduling algorithm must be clairvoyant.*

Theorem 5.3 therefore shows that having a partial prediction of the future can help to construct a better schedule than that constructed by a fully non-clairvoyant scheduler such as EDF. Theorem 5.4 further establishes a lower bound on the omniscience of the scheduler required to construct an energy-valid schedule [CHE 14c].

THEOREM 5.4.– *Let D be the greatest relative deadline of the application. No online ld-omniscient scheduling algorithm is optimal for the RTEH model if $ld < D$.*

Theorem 5.4 therefore gives us the clairvoyance horizon required by any optimal scheduler. The value of the greatest relative deadline appears to be a key parameter of the application. If we cannot predict the profile of the energy harvested over an interval of duration greater than or equal to this relative deadline, it is unreasonable to expect to achieve optimal online scheduling.

Estimating the quantity of energy extracted from the surroundings over a certain time interval is therefore one of the most important design questions of RTEH systems. The ambient energy source can be formally modeled or even precisely determined offline for some applications. But when it is uncontrollable and highly unstable, online prediction techniques applied cyclically over a moving time window are required to obtain lower bounds for the future harvested energy.

4.4. Concepts

4.4.1. *Concepts related to time*

Recall that the processor demand on $[t_1, t_2)$, which is written here as $tdbf(\mathcal{T}, t_1, t_2)$ (time demand bound function), is defined by the amount of processing time required by all jobs with activation times at or after t_1 and deadlines at or before t_2:

$$tdbf(\mathcal{T}, t_1, t_2) = \sum_{t_1 \leq r_k,\, d_k \leq t_2} C_k \qquad [4.1]$$

If the jobs in the set $\mathcal{T}$ under-utilize the processor, there is remaining available processor time and some temporal flexibility in the times when the jobs can be executed, which leads to the idea of *slack time*. Schedulability tests for EDF based on the processor demand approach calculate this processor time on every time interval beginning with an activation time and ending with a deadline. For each interval, it is checked that processing overload does not occur. This approach is equivalent to calculating the *static slack time*, denoted as $SST_{\mathcal{T}}(t_1, t_2)$ and given by:

$$SST_{\mathcal{T}}(t_1, t_2) = t_2 - t_1 - tdbf(\mathcal{T}, t_1, t_2) \qquad [4.2]$$

$SST_{\mathcal{T}}(t_1, t_2)$ corresponds to the duration of the maximal interval contained in $[t_1, t_2)$ during which the processor can remain idle while still guaranteeing the execution of the jobs in $\mathcal{T}$ with activation times at or after t_1 and deadlines at or before t_2. From this, we deduce the static slack time of the set $\mathcal{T}$:

$$SST_{\mathcal{T}} = \min_{0 \leq t_1 < t_2 \leq d_{Max}} SST_{\mathcal{T}}(t_1, t_2) \qquad [4.3]$$

Showing that $SST_{\mathcal{T}} \geq 0$ amounts to proving that the set $\mathcal{T}$ is schedulable by EDF in the absence of energy constraints.

For applications in which the jobs arrive at unpredictable times, schedulability analysis takes the form of an online test (we call this an admission test) that decides whether to accept or reject the new job [BUT 05]. Denote t_c the current time in the sequence constructed for $\mathcal{T}$ by a given scheduling algorithm. Let AT_i be the remaining execution time at t_c of unfinished jobs with deadlines at or before d_i. The slack time of the job τ_i at the current time t_c is given by:

$$ST_{\tau_i}(t_c) = d_i - t_c - tdbf(\mathcal{T}, t_c, d_i) - AT_i \qquad [4.4]$$

The quantity $ST_{\tau_i}(t_c)$ is the total quantity of processor time available in $[t_c, d_i)$ after finishing the execution of all jobs with deadlines at or before d_i. We can define the slack time of $\mathcal{T}$ at the current time t_c as follows:

$$ST_{\mathcal{T}}(t_c) = \min_{d_i > t_c} ST_{\tau_i}(t_c) \qquad [4.5]$$

The slack time calculated by [4.5] is the non-discontinuous slack time after t_c during which the processor can remain idle or execute other jobs than those in the set $\mathcal{T}$. Calculating $ST_{\mathcal{T}}(t_c)$ requires the construction of the EDL schedule starting from t_c, as was originally described in [CHE 89].

4.4.2. *Concepts related to energy*

Here, we will introduce new concepts in order to analyze the feasibility of a set of jobs that are additionally characterized by their energy requirements. Denote $E_p(t_1, t_2)$ the quantity of energy harvested between times t_1 and t_2. The energy demand (energy demand bound function) of a set of jobs $\mathcal{T}$ on the interval $[t_1, t_2)$, written as $edbf(\mathcal{T}, t_1, t_2)$, is given by:

$$edbf(\mathcal{T}, t_1, t_2) = \sum_{t_1 \leq r_k,\, d_k \leq t_2} E_k \qquad [4.6]$$

We can thus define the static slack energy of a set of jobs $\mathcal{T}$ on the interval $[t_1, t_2)$ by:

$$SSE_{\mathcal{T}}(t_1, t_2) = C + E_p(t_1, t_2) - edbf(\mathcal{T}, t_1, t_2) \qquad [4.7]$$

$SSE_{\mathcal{T}}(t_1, t_2)$ is the maximum quantity of energy available on the interval $[t_1, t_2)$ while guaranteeing the execution of the jobs of $\mathcal{T}$ with activation times after t_1 and deadlines at or before t_2. We can then define the static slack energy of $\mathcal{T}$ as follows:

$$SSE_{\mathcal{T}} = \min_{0 \leq t_1 < t_2 \leq d_{Max}} SSE_{\mathcal{T}}(t_1, t_2) \qquad [4.8]$$

The static slack energy of $\mathcal{T}$ is the energy surplus that can be consumed at any moment while still guaranteeing that the energy requirements of the jobs of $\mathcal{T}$ will be satisfied.

Consider the state of the system at the current time t_c. The slack energy of the job τ_i at t_c is given by:

$$SE_{\tau_i}(t_c) = E(t_c) + E_p(t_c, d_i) - edbf(\tau_i, t_c, d_i) \qquad [4.9]$$

$SE_{\tau_i}(t_c)$ is the maximum quantity of energy that can be consumed in $[t_c, d_i)$ while still guaranteeing that the energy requirements of jobs activated after t_c and with deadlines at or before d_i will still be met. If there is a job τ_i, such that $SE_{\tau_i}(t_c) = 0$, then executing any job with a deadline after d_i between t_c and d_i will cause an energy shortage for τ_i.

Based on these definitions, we can now introduce a new scheduler ED-H that is optimal for the RTEH model.

4.5. The ED-H scheduler

4.5.1. *Principles*

The underlying idea of the ED-H scheduler is to execute jobs according to the principle of EDF over time intervals determined as a function of energy constraints. Jobs are only executed after checking that performing one unit of processing will not cause an energy shortage for that job or any other job at some point in the future. ED-H is not equivalent to EDS, nor to EDL. This scheduler considers both the time and energy properties of jobs as well as the charge rate of the storage unit in order to decide whether to place the processor on standby. Broadly, ED-H is a variant of EDF that may be described as energy aware, as it has the ability to avoid energy shortages. Classical EDF is described as greedy because it systematically executes all jobs as early as possible, thus depleting the stored energy without regard for future energy requirements. Consider a set of jobs that is time-feasible by EDF. Energy shortage for a job τ_i can only occur if a job τ_j is executed before the arrival of τ_i, such that $d_j > d_i$. Indeed, the energy shortage for τ_i caused by τ_j, such that $d_j \leq d_i$, could not be avoided by any scheduler. Intuitively, some clairvoyance of the arrival of jobs and the energy production will allow EDF to anticipate a potential energy shortage and consequently avoid a missed deadline. The basic principle of ED-H is to authorize the execution of jobs whenever energy shortage cannot occur. This leads us to introduce the notion of *preemption slack energy* at the current time t_c as the largest quantity of energy that can be consumed by the active job without compromising the feasibility of any jobs that could preempt it.

Let d be the deadline of the active job at time t_c. The preemption slack energy of the set $\mathcal{T}$ at t_c is given by:

$$PSE_{\mathcal{T}}(t_c) = \min_{t_c < r_i < d_i < d} SE_{\tau_i}(t_c) \qquad [4.10]$$

4.5.2. *Algorithm specification*

Denote $L_r(t_c)$ as the list of jobs ready at time t_c. The ED-H scheduler consists of more than just a single rule for choosing a job ready for execution. It also includes a rule for dynamically managing the processor activity by specifying the intervals in which the processor should remain on standby, and those on which it should execute a job:

– Rule 1: The priorities assigned by EDF are used to select the next ready job in $L_r(t_c)$;

– Rule 2: The processor is on standby during $[t_c, t_c + 1)$ if $L_r(t_c) = \emptyset$;

– Rule 3: The processor is on standby during $[t_c, t_c + 1)$ if $L_r(t_c) \neq \emptyset$ and one of the following conditions is true:

1) $E(t_c) \approx 0$

2) $PSE_{\mathcal{T}}(t_c) \approx 0$;

– Rule 4: The processor is busy during $[t_c, t_c + 1)$ if $L_r(t_c) \neq \emptyset$ and one of the following conditions is true:

1) $E(t_c) \approx C$

2) $ST_{\mathcal{T}}(t_c) = 0$;

– Rule 5: The processor can be either busy or on standby if $L_r(t_c) \neq \emptyset$, $0 < E(t_c) < C$, $ST_{\mathcal{T}}(t_c) > 0$ and $PSE_{\mathcal{T}}(t_c) > 0$.

Rule 3 states that no job may be executed if the storage unit is empty, or if execution would lead to an unavoidable energy shortage due to insufficient preemption slack energy. Rule 4 indicates that the processor cannot remain idle if the storage unit is at full capacity, or if idling would cause a missed deadline as a result of zero slack time. If the storage unit is neither empty nor full, and if the system has non-zero slack time and non-zero preemption slack energy, Rule 5 states that the processor can be either on standby or busy without affecting the validity of the subsequent schedule. Note that with ED-H, energy is only wasted if the storage unit is full and no jobs are ready.

This description of ED-H does not mention the special case where the storage unit is full ($C \leq E(t_c) < C + e_{Max}$) with zero preemption slack energy ($0 \leq PSE_{\mathcal{T}}(t_c) < e_{Max}$). To avoid wasting energy by placing the processor on standby, the highest-priority job on $[t_c, t_c + 1)$ can be dispatched, incurring an energy consumption of at most e_{Max} units. The processor then remains idle until the storage unit has fully recharged. Thus, ED-H repeatedly switches between busy and standby so that over this period, the energy consumption is equal to the energy production. This wastes at most e_{Max} units of energy.

A number of different implementations can be derived for ED-H from the choice in Rule 5. ASAP and ALAP are special cases, which amount to executing jobs either systematically as early as possible when there is sufficient energy to do so, or as late as possible without allowing the storage unit to exceed full capacity. The choice of rule for deciding when to begin and when to end the storage charging phase determines the variant of ED-H. For example, we could choose to execute jobs whenever the energy level exceeds a certain threshold, and place the processor on standby to recharge the storage unit whenever the charge level is below another prespecified threshold. The implementation of the ED-H scheduler is therefore highly flexible, subject to the condition that at no point in time should any energy be wasted and that the system should be notified of any negative slack time or preemption slack energy.

4.5.3. *Performance*

It has been formally shown that ED-H is optimal [CHE 14a]. If ED-H cannot construct a valid schedule for a set of jobs $\mathcal{T}$ with a given configuration of hardware and energy conditions, then no other scheduler can.

The proof of optimality is based on the observation that job deadlines are missed either due to time shortage or energy shortage. In other words, the ED-H scheduler yields a valid schedule if there does not exist a time interval on which the processor demand exceeds the size of the interval and the energy demand exceeds the total energy available on the interval.

THEOREM 5.5.– *The ED-H scheduling algorithm is optimal for the RTEH model.*

The ED-H scheduler provides an optimal solution that is less restrictive than the LSA (Lazy Scheduling Algorithm) originally proposed in [MOS 07], a brief description of which will be given in the next section. LSA assumes that the energy consumed by each job varies linearly with its execution time.

4.5.4. *Clairvoyance*

According to the result stated earlier, no online scheduling algorithm can produce an optimal solution without knowledge of future information, described as *clairvoyance*, over a period of at least D units of time. This means that in order to make a decision at any current time t_c, the ED-H scheduler requires knowledge of both the process of job arrivals and the process of energy production for D units of time after t_c. This leads us to theorem 5.6, which is formally proven in [CHE 14a].

THEOREM 5.6.– *The ED-H scheduler is D-omniscient.*

This prerequisite of omniscience is therefore a key technological aspect of the implementation of the ED-H scheduler. In general, the executed jobs comprising the processing load are generated by a set of periodic tasks. Their arrival times are perfectly known. For the energy, the challenge is to predictively estimate the energy harvested over a moving time window of D units of time. This will be solved using different prediction methods specific to the type and source of the regenerative energy [LIU 11].

4.5.5. *Schedulability condition*

In order to check the feasibility of an application, it must be determined whether a set of jobs $\mathcal{T}$ is schedulable by the ED-H scheduler. The goal is to formally verify that it is possible to construct a valid ED-H schedule for the chosen physical hardware, energy storage unit and energy harvesting unit. This feasibility test will therefore consider the properties of its components, in particular the capacity of the storage unit and the power production of the harvesting unit. Theorem 5.7 shows that we can consider time and energy constraints separately. The feasibility test is subdivided into a time feasibility test and an energy feasibility test. In other words, an application is feasible if and only if it is time-feasible and energy-feasible [CHE 14a].

THEOREM 5.7.– *A set of jobs $\mathcal{T}$ is feasible if and only if*

$$SST_{\mathcal{T}} \geq 0 \text{ and } SSE_{\mathcal{T}} \geq 0 \qquad [4.11]$$

The complexity of the time feasibility test is $O(n^2)$, as the static slack time must be calculated on n^2 different intervals. If the ambient energy can be predictively estimated on each time interval by a finite number of values, then the energy feasibility test also has a computational complexity of $O(n^2)$.

4.5.6. *Illustrative example*

Consider once again the previous example. We will apply theorem 5.7 to test the feasibility of the two jobs τ_1 and τ_2. Calculating the static slack time $SST_{\mathcal{T}}$ with equation [4.3] is equivalent in this case to evaluating the four following quantities: $SST_{\mathcal{T}}(0,3)$, $SST_{\mathcal{T}}(0,8)$, $SST_{\mathcal{T}}(2,3)$ and $SST_{\mathcal{T}}(2,8)$. These are respectively equal to 2, 4, 0 and 5. Hence $SST_{\mathcal{T}} = 0$, which proves that there exists a time-valid schedule for the set $\mathcal{T}$. Calculating the static slack energy $SSE_{\mathcal{T}}$ with equation [4.10] is equivalent to evaluating the four following quantities: $SSE_{\mathcal{T}}(0,3)$, $SSE_{\mathcal{T}}(0,8)$, $SSE_{\tau}(2,3)$ and $SSE_{\mathcal{T}}(2,8)$. These are respectively equal to 8, 0, 4 and 14. Hence $SSE_{\mathcal{T}} = 0$, which proves that there exists an energy-valid schedule for the set $\mathcal{T}$. Finally, we deduce that the set $\mathcal{T}$ is guaranteed schedulable by ED-H.

We will now construct the ED-H schedule. Equation [4.9] gives $SE_{\tau_2}(0) = E(0) + E_p(0, d_2) - edbf(\tau_2, 0, d_2) = 8 + 6 - 6 = 8$. $SE_{\tau_2}(0)$ is the maximum amount of energy that can be consumed by any executed job starting from time zero, such that τ_2 remains feasible. τ_1, which is allowed to consume at most eight units of energy, begins execution and ends at time 1 since $SE_{\tau_2}(1) \approx 0$. Rule 3 places the processor on standby to recharge the storage unit in order to satisfy the future needs of τ_2. At time 2, τ_2 is activated and $E(2) = 6$. τ_2, which is higher priority than τ_1, has sufficient energy and executes fully. At time 3, $E(3) = 2$. Rule 3 places the processor on standby. The maximum duration of the standby can be calculated as follows. $ST_{\mathcal{T}}(3)$ is calculated, i.e. $ST_{\tau_1}(3)$ as τ_1 is the only job of the application. $ST_{\tau_1}(3) = d_1 - 3 - AT_1$, which is equal to three units of time. We can therefore decide to leave the processor on standby until time 6, unless the storage unit reaches the authorized charge threshold before time 6. In this example, the energy storage unit reaches full capacity at the same time that the slack time becomes zero. τ_1 is then executed from time 6 to 8, ending exactly on its deadline with an empty energy storage unit (see Figure 4.4).

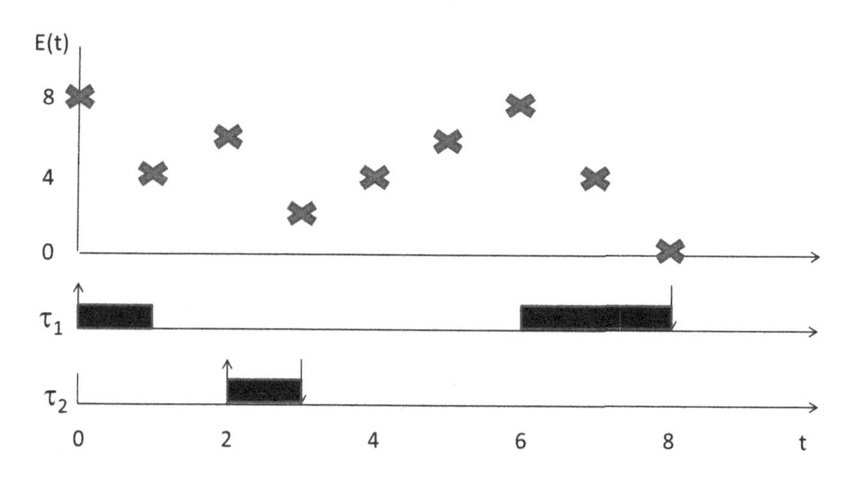

Figure 4.4. *Schedule constructed by the* ED-H *scheduler*

4.6. Another scheduling solution: LSA

The *LSA (Lazy Scheduling Algorithm)* was described in 2006 by Moser *et al.* from ETH Zurich, and a proof of its optimality was given in [MOS 07]. LSA is a dynamic-priority scheduler since it uses the same rule as EDF. Like ED-H, it differs from EDF in its policy for dynamically managing the processor activity.

4.6.1. *Hypotheses*

The optimality of LSA assumes that the processing system has the same instantaneous power consumption for all executed jobs. Thus, the total energy consumed by a job is proportional to its execution time and the coefficient of proportionality is given by the maximum power consumption of the processor. We therefore have that $C_i = E_i/P_{max}$. With LSA, any job τ_i can be executed either with an instantaneous power consumption equal to the maximum power of the processor P_{max} ($P_c(t) = P_{max}$) or with a variable power consumption equal to the power of the regenerative energy source ($P_c(t) = P_p(t)$). The model therefore assumes an ideal processor that is capable of dynamically adapting its instantaneous consumption of power to the production power.

4.6.2. *Principles*

When a job τ_i arrives, the scheduler begins by calculating a start time for this job, written s_i. The processor is deliberately placed on standby between the arrival time and the start time in order to allow the storage unit to charge as much as possible, or potentially to consume energy at the same rhythm at which it is produced in the event that the storage unit reaches full capacity. This charging interval is calculated to ensure that the processor has sufficient energy to operate continuously at power P_{max} in $[s_i, d_i]$ with no risk of energy shortage, but also such that all available energy is consumed at time d_i, so that $E_c(d_i) = 0$.

To find the optimal value for s_i, the total available energy in the time interval $[r_i, d_i]$ is calculated as $E_c(r_i) + E_p(r_i, d_i)$. Then, the minimum period required to consume this energy without interruptions is calculated using the equation: $\frac{E_c(r_i) + E_p(r_i, d_i)}{P_{max}}$. The optimal time to start the execution of the job τ_i may then be deduced as follows:

$$s_i^* = d_i - \frac{E_c(r_i) + E_p(r_i, d_i)}{P_{max}} \qquad [4.12]$$

Note, however, that we must also consider the case where the available energy is exhausted at some time in the interval $[s_i^*, d_i]$. The start time $s_i^{'}$ of the job τ_i is calculated as follows:

$$s_i^{'} = d_i - \frac{C + E_p(s_i^{'}, d_i)}{P_{max}} \qquad [4.13]$$

It follows that the optimal start time s_i of the job τ_i is given by the larger of the two above quantities:

$$s_i = max(s_i^*, s_i^{'}) \qquad [4.14]$$

In summary, the LSA scheduler is implemented by the following rules:

– Rule 1: The priority assigned by EDF is used to select the next ready task in $L_r(t_c)$;

– Rule 2: The start time s_i of each job τ_i is calculated using Formula 4.14;

– Rule 3: The processor is on standby during $[t_c, t_c + 1)$ if $L_r(t_c) = \emptyset$;

– Rule 4: The processor is on standby during $[t_c, t_c + 1)$ if $L_r(t_c) \neq \emptyset$ and $t_c < s_i$;

– Rule 5: The processor is busy during $[t_c, t_c + 1)$ with power $P_c(t) = P_{max}$ if $t_c \geq s_i$;

– Rule 6: The processor is busy during $[t_c, t_c + 1)$ with power $P_c(t) = P_p(t)$ if $t_c < s_i$ and $E(t_c) \approx C$.

4.6.3. *Schedulability condition*

THEOREM 5.8.– *A set of jobs $\mathcal{T}$ is schedulable by LSA if and only if*

$$edbf(t_1, t_2) \leq min(E_p(t_1, t_2) + C, P_{max} \cdot (t_2 - t_1)), \quad \forall t_1, t_2 \geq 0 \qquad [4.15]$$

Observe that the schedulability test given by equation [4.15] may be reduced to the schedulability test of ED-H given by equation [4.11], as $edbf(t_1, t_2) = P_{max} \times tdbf(t_1, t_2)$. The condition $edbf(t_1, t_2) \leq$

$P_{max} \cdot (t_2 - t_1)$ means that the processor demand of the set of jobs activated after t_1 with deadlines at or before t_2 must be less than the size of the interval $[t_1, t_2)$, which is (t_2-t_1). This condition would be sufficient to test the feasibility of the system in the absence of energy constraints.

The condition $edbf(t_1, t_2) \leq E_p(t_1, t_2) + C$ means that the maximum quantity of energy required by the set of jobs between t_1 and t_2 must be less than or equal to the total quantity of energy produced by the source between t_1 and t_2 plus the maximum quantity of energy available in the storage unit, which is C.

4.6.4. *Illustrative example*

Consider again the previous example. This was a special case in which the total energy consumed by each job is proportional to its execution time. The LSA scheduler may therefore be applied. At time 0, the activation time of τ_1, we can use equation [4.12] to calculate s_1^*. $s_1^* = d_1 - \dfrac{E_c(r_1) + E_p(r_1, d_1)}{P_{max}}$, with $P_p = 2$ $P_{max} = 6$, $E_c(r_1) = 8$, $r_1 = 0$ and $d_1 = 8$. Hence $s_1^* = 4$. Returning to equation 4.13, we must find $s_1^{'}$, such that $P_{max} \times (d_1 - s_1^{'}) = C + E_p(s_1^{'}, d_1)$ with $E_p(s_1^{'}, d_1) = P_p \times d_1 - P_p \times s_1^{'}$.

This leads to the equation: $48 - 6s_1^{'} - 8 - 16 - 2s_1^{'} = 0$. This implies $8s_1^{'} = 24$, so $s_1^{'} = 3$. Since $s_1 = max(s_1^*, s_1^{'})$, we deduce that the start time of τ_1 is $s_1 = 4$. According to Rule 6, the processor is busy between time 0 and time 2 with a power consumption equal to the power production, which is two units energy per unit time. At time 2, the storage unit contains $E_c(2) = 6$. The activation of τ_2 at time 2 leads to the calculation of s_2. $s_2^* = d_2 - \dfrac{E_c(r_2) + E_p(r_2, d_2)}{P_{max}}$, so $s_2^* = 3 - \dfrac{6+2}{6}$ and thus $s_2^* = \dfrac{8}{6}$. As before, $P_{max} \times (d_2 - s_2^{'}) = C + P_p \times d_2 - P_p \times s_2^{'}$. This gives us $18 - 6s_2^{'} - 8 - 6 - 2s_2^{'} = 0$. Hence $8s_2^{'} = 4$, and thus $s_2^{'} \approx 0$. It follows that $s_2 = 2$ which means that τ_2 is executed immediately until time 3. As a result of Rule 4, the processor is placed on standby until time $s_1 = 4$, at which point the processor executes τ_1 and finishes execution on the deadline. Note that in this example, the schedule produced by LSA is identical to the schedule constructed by ED-H.

4.7. Technological hurdles

The implementation of an energy-aware scheduler such as *ED-H* or *LSA* requires a prediction of the energy supply, dynamic measurements of the level of stored energy and a feasibility test. Currently, no RTOS provides these features.

4.7.1. *Estimation of the harvested energy*

The amount of harvested energy varies over time in a non-deterministic manner. In a classical system powered by energy delivered by a battery, simply measuring the remaining capacity of the battery gives the amount of available energy. In energy harvesting systems, monitoring the residual capacity is not enough. Additional sophisticated methods are required to characterize the energy source. The literature provides methods for predicting solar energy [KAN 07]. Most methods derive an estimate of the near-future harvested energy from the past harvested energy. For example, the technique of moving average considers the statistical average over a certain time window, continuously recalculated. The exponential moving average weights the terms with an exponential decay. Each data point for the power considered in the average is weighted by a certain factor more than the previous data point. Thus, more recent observations are viewed as more significant, without completely disregarding older data. The experiments described, in particular, by [LIU 11] highlight the significance of the choice of prediction technique on the resulting performance of the scheduler, evaluated in terms of met deadlines. Note that the optimality of the LSA and ED-H schedulers was formally shown subject to the assumption that a perfect prediction of the harvested energy is available.

4.7.2. *Measurement of stored energy*

The LSA and ED-H schedulers both assume knowledge of the current energy level in the storage unit, called the residual capacity. For ED-H, this quantity is used to calculate the slack energy of each job. Furthermore, the energy level is regularly measured in order to avoid any shortage during the execution of a job, or to avoid energy wastage when the system is on standby. Any valid implementation of ED-H or LSA assumes that this measurement is available with a precision of at least e_{Max}, which is the largest amount of energy that a job can consume in a single unit of time.

4.7.3. *Schedulability test*

The purpose of the schedulability test is to ensure that the physical hardware and the chosen source of ambient energy will allow the specified operational constraints to be observed, which in this case are given by the deadlines. The physical hardware consists of three main components:

– the processor, whose processing speed determines the execution time of each job. Consequently, this affects the slack time of each job (see equation [4.2]) for ED-H and the start time of each job for LSA;

– the energy storage unit. Its capacity determines, in particular, the size of the time interval during which the system can continue to operate without harvesting energy from the environment;

– the energy harvesting unit. We consider examples of uncontrollable energy sources that cannot be configured. The production power is determined solely by the size of the harvesting unit. The production power affects the value of the slack energy of each job (see equation [4.9]) for ED-H and the starting time for LSA (see equation [4.14]).

In classical real-time systems, the schedulability tests are conducted offline and are based on the fully deterministic quantities of the job execution times. The jobs are then scheduled online with a guarantee of feasibility, having considered the worst-case scenario. Recall also that the grand majority of real-time systems, whether autonomous or not, execute jobs generated by periodic tasks. As we saw in Chapter 2, the schedule constructed by any scheduler is repetitive. It is sufficient to perform the test on a single period of repetition, which reduces the complexity of the test implementation.

In the case of energetically self-sufficient systems, the test can only be performed offline if the energy production profile is perfectly known over the full lifetime of the application. For example, it can be assumed that the indoor solar energy in a building that is always on provides constant power over time. The energy production profile can also be cyclical. Outdoor solar energy has a period of 24 h. The test can therefore rely on the presence of a cycle to verify the energy neutrality of the system. In every other case, dynamically implemented feasibility tests over moving time windows are required, based on techniques such as those described above.

For ED-H, the schedulability test is performed in $O(n^2)$ operations using equations [4.2] and [4.7], where n is the number of jobs on the test interval. Note that this number of operations can be reduced by means of simple offline pre-calculation and online updating as described in [CHE 89] and [CHE 99]. The test may reach a negative conclusion regarding schedulability, namely that there is insufficient energy to execute the set of jobs without missing a deadline. This situation can then be managed in the same way as in classical real-time systems, for example by deleting certain jobs in order to reduce the processing load.

4.8. Conclusion

There are a number of problems that must be solved when designing an autonomous real-time system, relating to the harvesting, storage and utilization of ambient energy. The goal is to ensure long-term self-sufficiency while guaranteeing acceptable levels of real-time performance. Energy-neutral operation is defined by the capacity of the system to meet all of its operational constraints using only the energy available in its storage system without ever experiencing an energy shortage. In this chapter, we restricted attention to the case of uniprocessor, monofrequency platforms. We presented an optimal scheduler, ED-H, which is a variant of the Earliest Deadline First scheduler. The most significant difference with EDF lies in the dynamic specification of the intervals during which the processor is placed on standby in order to prevent excessive energy consumption and ensuing energy shortage. ED-H provides not only a real-time scheduling policy, but also a dynamic power management strategy. We emphasized, in particular, that this technique only achieves best performance in ideal operational conditions. Indeed, if the energy source is fully probabilistic in nature and estimating the future energy is impossible, the classical EDF scheduler remains the best choice.

Energetically self-sufficient real-time systems have also been studied in the context of platforms with DVFS features [LIU 08, LIU 09, LIU 11], fixed-priority systems [ABD 13] and multiprocessor platforms [LU 11].

Conclusion

In this book, we examined the problem of real-time scheduling in systems powered by a regenerative energy source. These systems harvest energy from their surroundings, store it in energy storage units such as batteries and/or supercapacitors and then use it to power themselves. This technology is increasingly used in small embedded systems that are required to operate continuously over long periods of time. A typical example is given by wireless sensor networks and medical implants. Furthermore, most of the processing performed by these autonomous systems is subject to real-time execution constraints. It is crucial that the results produced are not just logically correct, but are delivered within a certain time window.

This book focused attention on real-time scheduling designed to achieve energy self-sufficiency on uniprocessor platforms. In this context, the objective is to establish suitable real-time scheduling policies that can consider the additional dimension of energy, and that can do so dynamically. As well as specifying order relations between concurrent jobs, a strategy for dynamic management of the processor activity must be chosen. Any future missed deadline due to either processing overload or energy shortage must be anticipated in order to minimize the impact on the system. This requires choosing a suitable size for the storage unit (e.g. battery) and harvesting unit (e.g. solar panel) in order to guarantee the feasibility of the system in terms of energy.

After briefly introducing the subject matter of the book, in Chapter 1, we presented the principles and underlying concepts of real-time computing by means of a series of typical examples from avionics, multimedia and the medical sector. We gave a list of real-time operating systems adapted to embedded devices. We highlighted the fact that few of these operating systems offer energy-saving features, and that none are suitable in the context of an intermittent power supply; existing mechanisms reduce the power consumed by the electronic circuits and do not attempt to achieve energy-neutral operation from an uncontrollable ambient power source.

Chapter 2 established the fundamental principles of real-time scheduling and recalled the most significant scientific results regarding uniprocessor architectures. In Chapter 3, we gave a description of the most prominent forms of energy present in the environment, with a particular focus on sources of energy suitable for use by small embedded systems such as so-called smart objects. Finally, in Chapter 4, we showed how an autonomous real-time system can be modeled by simultaneously describing the energy requirements of real-time tasks in an application and the energy generated by the environment. On the basis of this model, we introduced a new optimal scheduler, based on the same priority rules as Earliest Deadline First, but relying on predictions of the harvested energy.

In summary, this book elucidates the stakes and the difficulties associated with the implementation of autonomous real-time systems. This is a relatively recent field of research, and a number of challenges must still be addressed to efficiently and simultaneously manage the two dimensions of *time* and *energy*. In particular, the most appropriate approach to modeling the system as a whole could be examined. *Which assumptions should we make in order to model the energy consumption profiles of application tasks as best as possible (given that the consumption of any task varies over time as a function of the different circuits involved in its processing)? How can we integrate potential synchronization constraints associated with shared access to resources?* At the same time, processors are continuously evolving and implement an increasing array of energy-related techniques. One example of this is the technique of *Dynamic Voltage and Frequency Scaling (DVFS)* in order to reduce (or increase) the processor frequency, which improves the energy consumption, but negatively impacts the execution time. *How can we optimally exploit this kind of processor to guarantee the observance of time and energy constraints in autonomous real-time systems?* The scientific community must find an answer to each of these questions in the near future.

Bibliography

[ABD 13] ABDEDDAIM Y., CHANDARLI Y., MASSON D., "The optimality of PFPasap algorithm for fixed-priority energy-harvesting real-time systems", *25th Euromicro Conference on Real-Time Systems*, 2013.

[ABE 98] ABENI L., BUTTAZZO G., "Integrating multimedia applications in hardware real-time systems", *Proceedings of the 19th IEEE Real-Time Systems Symposium*, pp. 4–14, 1998.

[AHO 74] AHO A.-V., HOPCROFT J.-E., ULLMAN J.-D., *The Design and Analysis of Computer Algorithms*, Addison-Wesley, Reading, 1974.

[AKY 02] AKYILDIZ I.F., SU W., SANKARASUBRAMANIAM Y. *et al.*, "A survey on sensor networks", *Communications Magazine, IEEE*, vol. 40, no. 8, pp. 102–114, 2002.

[AKY 10] AKYILDIZ I., VURAN M.C., *Wireless Sensor Networks*, John Wiley & Sons, New York, 2010.

[ALL 01] ALLAVENA A., MOSSE D., "Frame-based embedded systems with rechargeable batteries", *Workshop on Power Management for Real-Time and Embedded Systems*, 2001.

[AUD 90] AUDSLEY N., BURNS A., "Deadline-monotonic scheduling", Technical Report YCS146, Department of Computer Science, University of York, 1990.

[AUD 91] AUDSLEY N.-C., BURNS A., RICHARDSON M. *et al.*, "Hard real-time scheduling: the deadline monotonic approach", *Proceedings of the 8th IEEE Workshop on Real-Time Operating Systems and Software*, Atlanta, 1991.

[AUD 93] AUDSLEY N.-C., BURNS A., RICHARDSON M. *et al.*, "Applying new scheduling theory to static priority preemptive scheduling", *Software Engineering Journal*, vol. 8, no. 5, pp. 284–292, 1993.

[BAR 90] BARUAH S.-K., HOWELL R.-R., ROSIER L.-E., "Algorithms and complexity concerning the preemptive scheduling of periodic real-time tasks on one processor", *Real-Time Systems*, vol. 2, no. 4, pp. 301–324, 1990.

[BAR 91] BARUAH S., KOREN G., MISHRA B. *et al.*, "Online scheduling in the presence of overload", *Symposium on Foundations of Computer Science*, pp. 100–110, 1991.

[BAR 92] BARUAH S., KOREN G., MAO D. *et al.*, "On the competitiveness of on-line real-time job scheduling", *Real-Time Systems*, vol. 4, no. 2, pp. 125–144, 1992.

[BAR 93] BARUAH S.-K., HOWELL R.-R., ROSIER L.-E., "Feasibility problems for recurring tasks on one processor", *Theoretical Computer Science*, vol. 118, pp. 3–20, 1993.

[BAR 12] BARRY P., CROWLEY P., *Modern Embedded Computing: Designing Connected, Pervasive, Media-Rich Systems*, Morgan Kaufmann Publishers Inc., San Francisco, 2012.

[BAT 98] BATES I.-J., Scheduling and timing analysis for safety critical real-time systems, PhD Thesis, University of York, 1998.

[BER 01] BERNAT G., BURNS A., LLAMOSI A., "Weakly-hard real-time systems", *IEEE Transactions on Computers*, vol. 50, no. 4, pp. 308–321, 2001.

[BIN 01] BINI E., BUTTAZZO G.-C., BUTTAZZO G.-M., "A hyperbolic bound for the rate monotonic algorithm", *Proceedings of the 13th Euromicro Conference on Real-Time Systems*, IEEE 2001, Delft, The Netherlands, pp. 59–66.

[BLA 76] BLAZEWICZ J., "Deadline scheduling of tasks – a survey", *Foundations of Control Engineering*, vol. 1, no. 4, pp. 203–216, 1976.

[BON 99] BONNET C., DEMEURE I., *Introduction aux systmes temps réel*, Hermes Science, Paris, p. 207, 1999.

[BOU 91] BOUCHENTOUF T., Ordonnancement sous contraintes de précédence dans les systèmes temps-réel, PhD Thesis, University of Nantes, 1991.

[BOV 05] BOVET D.-P., CESATI M., *Understanding the Linux Kernel*, O'Reilly, 3rd edition, 2005.

[BUT 97] BUTTAZZO G.-C., *Hard Real-Time Computing Systems*, Kluwer Academic, 1997.

[BUT 99] BUTTAZZO G.-C., SENSINI F., "Optimal deadline assignment for scheduling soft aperiodic tasks in hard real-time environments", *IEEE Transactions on Software Engineering*, vol. 25, no. 1, pp. 22–32, 1999.

[BUT 05] BUTTAZZO G.-C., *Hard Real-Time Computing Systems: Predictable Scheduling Algorithms and Applications*, Springer, Berlin, 2005.

[CAR 13] CAROFF T., ROUVIÈRE E., WILLEMIN J., "Thermal energy harvesting", *Energy Autonomous Micro and Nano Systems*, pp. 153–184, Wiley Online Library, 2013.

[CHA 08] CHALASANI S., CONRAD J.M., "A survey of energy harvesting sources for embedded systems", *IEEE Southeastcon 2008*, IEEE, pp. 442–447, 2008.

[CHA 10] CHAOUCHI H. (ed.), *The Internet of Things: Connecting Objects*, ISTE Ltd, London and John Wiley & Sons, New York, 2010.

[CHE 89] CHETTO H., CHETTO M., "Some results of the earliest deadline scheduling algorithm", *IEEE Transactions on Software Engineering*, vol. 15, no. 10, pp. 1261–1269, 1989.

[CHE 99] CHETTO-SILLY M., "The EDL server for scheduling periodic and soft aperiodic tasks with resource constraints", *Real-Time Systems*, vol. 17, no. 1, pp. 87–111, 1999.

[CHE 14a] CHETTO M., "Optimal scheduling for real-time jobs in energy harvesting computing systems", *IEEE Transactions on Emerging Topics in Computing*, vol. 2, no. 2, pp. 122–133, 2014.

[CHE 14b] CHETTO M., QUEUDET A., "A note on EDF scheduling for real-time energy harvesting systems", *IEEE Transactions on Computers*, vol. 63, no. 4, pp. 1037–1040, 2014.

[CHE 14c] CHETTO M., QUEUDET A., "Clairvoyance and online scheduling in real-time energy harvesting systems", *Real-Time Systems*, vol. 50, no. 2, pp. 179–184, 2014.

[COT 00] COTTET F., DELACROIX J., KAISER C. *et al.*, *Ordonnancement Temps-réel*, Hermes Science, Paris, 2000.

[DAV 93] DAVIS R.-I., TINDELL K.-W., BURNS A., "Scheduling slack time in fixed priority preemptive systems", *Proceedings of the Real-Time Systems Symposium*, pp. 222–231, 1993.

[DEC 02] DECOTIGNY D., "Introduction l'ordonnancement dans les systèmes temps-réel", Bibliographical report, IRISA, Rennes, 2002.

[DER 74] DERTOUZOS M.-L., "Control robotics: the procedural control of physical processes", in ROSENFELD J.L (ed), *Information Processing 74: Proceedings of IFIP Congress 74*, North Holland Publishing Company, 1974.

[DES 13] DESPESSE G., CHAILLOUT J.J., BOISSEAU S. *et al.*, "Mechanical energy harvesting", *Energy Autonomous Micro and Nano Systems*, pp. 115–151, Wiley Online Library, 2013.

[DIM 14] DIMITRAKOPOULOS N., "La diffusion Ultra HD 4K de bout en bout", available at: http://www.mediakwest.com/broadcast/workflow-cloud/item/diffusion-uhd-4k-de-bout-en-bout.html, 2014.

[DIS 16] "Disponen RTOS gratuito a PyMEs, estudiantes y escuelas", *Electronicos On-Line Magazine*, available at: http://www.electronicosonline.com/etiqueta/academia, 2016.

[DOR 91] DORSEUIL A., PILLOT P., *Le temps-réel en milieu industriel, concepts, environnements, multitâches*, Bordas, Paris, 1991.

[DOU 88] DOUGLAS AIRCRAFT COMPANY, "Advanced concept ejection seat ACES II", Report MDC J4576 Revision D, March 1988.

[FRE 16] FREERTOS+NABTO, "What is FreeRTOS+Nabto?", http://www.freertos.org/FreeRTOS-Plus/Nabto/what_is_freertos_plus_nabto.shtml, 2016.

[GEO 95] GEORGE L., MUHLETHALER P., RIVIERRE N., "Optimality and non-preemptive real-time scheduling revisited", Research report RR-2516, INRIA, 1995.

[GEO 96] GEORGE L., RIVIERRE N., SPURI M., "Preemptive and non-premmptive real-time uni-processor scheduling", Research report RR-2966, INRIA, 1996.

[GHA 95] GHAZALIE T.-M., BAKER T.-P., "Aperiodic servers in a deadline scheduling environment", *The Journal of Real-Time Systems*, vol. 9, pp. 21–36, 1995.

[GRA 79] GRAHAM R.-L., LAWLER E.-L., LENSTRA J.-K. *et al.*, "Optimization and approximation in deterministic sequencing and scheduling", *Annals of Discrete Mathematics*, vol. 5, pp. 287–326, 1979.

[GUN 14] GUNES V., PETER S., GIVARGIS T. *et al.*, "A survey on concepts, applications, and challenges in cyber-physical systems", *KSII Transactions on Internet and Information Systems (TIIS)*, vol. 12, no. 12, pp. 4242–4268, 2014.

[HAM 97] HAMDAOUI M., RAMANATHAN P., "Evaluating dynamic failure probability for streams with (m,k)-firm deadlines", *IEEE Transactions on Computers*, vol. 46, no. 12, pp. 1325–1337, 1997.

[HOR 74] HORN W.-A., "Some simple scheduling algorithms", *Naval Research Logistics Quarterly*, vol. 21, pp. 177–185, 1974.

[HOR 76] HOROWITZ E., SAHNI S., "Exact and approximate algorithms for non identical processors", *Journal of the Association for Computing Machinery*, vol. 23, pp. 317–327, 1976.

[JAC 55] JACKSON J.-R., "Scheduling a production line to minimize maximum tardiness", Research report 43, University of California, Los Angeles, 1955.

[JAY 06] JAYASEELAN R., MITRA T., LI X., "Estimating the worst-case energy consumption of embedded software", *12th IEEE Real-Time and Embedded Technology and Applications Symposium*, 2006.

[JOS 86] JOSEPH M., PANDYA P., "Finding response time in a real-time system", *The Computer Journal*, vol. 29, no. 5, pp. 390–395, 1986.

[KAN 06] KANSAL A., HSU J., "Harvesting aware power management for sensor networks", *Proceedings of ACM/IEEE Design Automation Conference*, pp. 651–656, 2006.

[KAN 07] KANSAL A., HSU J., ZAHEDI S. *et al.*, "Power management in energy harvesting sensor networks", *ACM Transactions on Embedded Computing Systems*, vol. 6, no. 4, pp. 46–61, 2007.

[KEI 14] KEITH-HYNES P., MIZE B., ROBERT A. *et al.*, "The diabetes assistant: a smartphone-based system for real-time control of blood glucose", *Electronics*, vol. 3, pp. 609–623, 2014.

[KIR 15] KIRK A.P., *Solar Photovoltaic Cells*, Academic Press, Oxford, 2015.

[KOP 91] KOPETZ H., "Event-triggered versus time-triggered real-time systems", in KARSHMER A., NEHMER J. (eds), *Proceedings of the Workshop on Operating Systems of the 90s and Beyond, Lecture Notes in Computer Science*, vol. 563, pp. 87–101, Springer-Verlag, 1991.

[LAG 76] LAGEWEG B.-J., LENSTRA J.-K., RINNOY KAN A.-H.-G., "Minimizing maximum lateness on one machine: computational experience and some applications", *Statistica Neerlandica*, vol. 30, no. 1, pp. 25–41, 1976.

[LAW 83] LAWLER E.-L., "Recent results in the theory of machine scheduling", in BACHEN A. *et al.*, (ed.) *Mathematical Programming: The State of the Art*, Springer-Verlag, New York, pp. 202–234, 1983.

[LE 10] LE SUEUR E., HEISER G., "Dynamic voltage and frequency scaling: the laws of diminishing returns", *Proceedings of the 2010 International Conference on Power Aware Computing and Systems*, HotPower'10, Berkeley, USENIX Association, pp. 1–8, 2010.

[LE 16] "Le Système Temps-Réel Déployé sur 3 milliards d'appareils", http://neomore.com/Nucleus_mentor.aspx, 2016.

[LEH 87] LEHOZCKY J.-P., SHA L., STROSNIDER K.-K., "Enhanced aperiodic responsivess in hard real-time environments", *Proceedings of the 13th IEEE Real-Time Systems Symposium*, pp. 261–270, 1987.

[LEH 89] LEHOZCKY J.-P., SHA L., DING Y., "The rate-monotonic scheduling algorithm: exact characterization and average case behaviour", *Proceedings of the IEEE Real-Time Systems Symposium*, pp. 166–171, 1989.

[LEH 92] LEHOZCKY J.-P., RAMOS-THUEL S., "An optimal algorithm for scheduling soft-aperiodic tasks in fixed-priority preemptive systems", *Proceedings of the 13th IEEE Real-Time Systems Symposium*, pp. 110–123, 1992.

[LEU 80] LEUNG J., MERRIL M., "A note on preemprive scheduling of periodic real-time tasks", *Information Processing Letters*, vol. 11, no. 3, pp. 115–118, 1980.

[LEU 82] LEUNG J.-Y.-T., WHITEHEAD J., "On the complexity of fixed-priority scheduling of periodic, real-time tasks", *Performance Evaluation*, vol. 2, pp. 237–250, 1982.

[LIU 73] LIU C.-L., LAYLAND J.-W., "Scheduling algorithms for multiprogramming in a hard real-time environnement", *Journal of the Association for Computer Machinery*, vol. 20, no. 1, pp. 46–61, 1973.

[LIU 08] LIU S., QIU Q., WU Q., "Energy aware dynamic voltage and frequency selection for real-time systems with energy harvesting", *Proceedings of the Conference on Design, Automation and Test in Europe*, pp. 236–241, 2008.

[LIU 09] LIU S., WU Q., QIU Q., "An adaptive scheduling and voltage/frequency selection algorithm for real-time energy harvesting systems", *ACM/IEEE Design Automation Conference*, pp. 782–787, 2009.

[LIU 11] LIU S., LU J., WU Q. *et al.*, "Harvesting-aware power management for real-time systems with renewable energy", *IEEE Transactions on Very Large Scale Integration (VLSI) Systems*, pp. 1–14, 2011.

[LIU 12] LIU R., *Electrochemical Technologies for Energy Storage and Conversion: Vol. 1*, Wiley-VCH, 2012.

[LOC 86] LOCKE C.-D., Best-effort decision making for real-time scheduling, PhD Thesis, Carnegie-Mellon University, 1986.

[LU 11] LU J., QIU Q., "Scheduling and mapping of periodic tasks on multi-core embedded systems with energy harvesting", *Conference on Green Computing*, pp. 1–6, 2011.

[MAS 04] MASMANO M., RIPOLL I., CRESPO A. *et al.*, "TLSF: a new dynamic memory allocator for real-time systems.", *Proceedings of ECRTS (2004)*, pp. 79–86, 2004.

[MCG 13] MCGRATH M.J., NI SCANAILL C. (eds), *Sensor Technologies: Healthcare, Wellness and Environmental Applications*, Apress, Berkeley, 2013.

[MOK 78] MOK A.-K., DERTOUZOS M.-L., "Multiprocessor scheduling in a hard real-time environment", *Proceedings of the 7th Texas Conference on Computer Systems*, pp. 76–81, 1978.

[MOK 83] MOK A.-K., Fundamental design problems of distributed systems for the hard real-time environment, PhD Thesis, MIT, 1983.

[MOO 03] MOORE D., Incorporating the SJU-17A naval aircrew common ejection seat in the EA-6B aircraft, Master's Thesis, University of Tennessee, 2003.

[MOS 07] MOSER C., BRUNELLI D., THIELE L. *et al.*, "Real-time scheduling for energy harvesting sensor nodes", *Real-Time Systems*, 2007.

[NUC 16] "Nucleus RTOS: power management APIs for low power design", https://www.mentor.com/embedded-software/nucleus/power-management, 2016.

[PAR 05] PARADISO J.A., STARNER T., "Energy scavenging for mobile and wireless electronics", *IEEE Pervasive Computing*, vol. 4, no. 1, pp. 18–27, 2005.

[PEN 11] PENELLA-LÓPEZ M.T., GASULLA-FORNER M., *Powering Autonomous Sensors: An Integral Approach with Focus on Solar and RF Energy Harvesting*, Springer Science & Business Media, 2011.

[PRI 09] PRIYA S., INMAN D.J., *Energy Harvesting Technologies*, Springer, 2009.

[RAG 05] RAGHUNATHAN V., KANSAL A., HSU J. *et al.*, "Design considerations for solar energy harvesting wireless embedded systems", *Proceedings of the 4th International Symposium on Information Processing in Sensor Networks*, IEEE Press, p. 64, 2005.

[RAN 04] RANDALL J., BHARATULA N., PERERA N. *et al.*, "Indoor tracking using solar cell powered system: interpolation of irradiance", *International Conference on Ubiquitous Computing*, 2004.

[RIC 05] RICHARD P., "Analyse du temps de réponse et de la demande processeur en ordonnancement temps-réel de tâches périodiques", *Ecole d'été Temps Réel (ETR'05)*, 2005.

[RID 05] RIDOUARD F., RICHARD P., COTTET F., "Ordonnancement de tâches indépendantes avec suspension", *Proceedings of the 13th International Conference on Real-Time Systems*, pp. 251–272, 2005.

[RIP 96] RIPOLL I., CRESPO A., MOK A.-K., "Improvement in feasibility testing for real-time tasks", *Journal of Real-Time Systems*, vol. 11, no. 1, pp. 19–39, 1996.

[SCH 79] SCHWARTZ M., Position and restraint system for aircrewman, US Patent 4437628, 1979.

[SER 72] SERLIN O., "Scheduling of time critical process", *Proceedings of the Joint Computer Conference*, vol. 40, pp. 925–932, 1972.

[SIL 93] SILLY-CHETTO M., "Sur la problématique de l'ordonnancement dans mes systèmes informatiques temps-réel", HDR, Université de Nantes, 1993.

[SIL 99] SILLY M., "The EDL server for scheduling periodic and soft aperiodic tasks with resource constraints", *The Journal of Real-Time Systems*, vol. 17, pp. 1–25, 1999.

[SIN 01] SINHA A., CHANDRAKASAN A., "Dynamic power management in wireless sensor networks", *Design & Test of Computers, IEEE*, vol. 18, no. 2, pp. 62–74, IEEE, 2001.

[SOD 05] SODANO H.A., INMAN D.J., PARK G., "Comparison of piezoelectric energy harvesting devices for recharging batteries", *Journal of Intelligent Material Systems and Structures*, vol. 16, no. 10, pp. 799–807, Sage Publications, 2005.

[SOR 74] SORENSON P.-G., A methodology for real-time system development, PhD Thesis, University of Toronto, 1974.

[SPR 89] SPRUNT B., SHA L., LEHOCZKY J.-P., "Aperiodic task scheduling for hard real-time systems", *The Journal of Real-Time Systems*, vol. 1, pp. 27–60, 1989.

[SPU 94] SPURI M., BUTTAZZO G.-C., "Efficient aperiodic service under earliest deadline scheduling", *Proceedings of the IEEE Real-Time Systems Symposium*, pp. 2–11, 1994.

[SPU 96a] SPURI M., Analysis of deadline schedule real-time systems, Technical Report 2772, INRIA, 1996.

[SPU 96b] SPURI M., BUTTAZZO G.-C., "Scheduling aperiodic tasks in dynamic priority systems", *The Journal of Real-Time Systems*, vol. 10, no. 2, 1996.

[STA 88] STANKOVIC J.-A., "A serious problem for next-generation systems", *IEEE Computer*, vol. 21, no. 10, pp. 10–19, 1988.

[STA 96] STARNER T., "Human-powered wearable computing", *IBM Systems Journal*, vol. 35, pp. 618–629, IBM, 1996.

[STR 95] STROSNIDER J.-K., LEHOCZKY J.-P., SHA L., "The deferrable server algorithm for enhanced aperiodic responsiveness in hard real-time environments", *IEEE Transactions on Computers*, vol. 44, no. 1, pp. 73–91, 1995.

[TIA 95] TIA T-S., LIU J.-W.-S., SHANKAR M., "Algorithms and optimality of scheduling aperiodic requests in fixed-priority preemptive systems", *The Journal of Real-Time Systems*, vol. 10, no. 1, pp. 23–43, 1995.

[VAN 13] VANGARI M., PRYOR T., JIANG L., "Supercapacitors: review of materials and fabrication methods", *Journal of Energy Engineering*, vol. 139, no. 2, pp. 72–79, 2013.

[WAL 11] WALTISPERGER G., Architectures intégrées de gestion de l'énergie pour les microsystèmes autonomes, PhD Thesis, University of Grenoble, 2011.

[ZHA 09] ZHANG F., BURNS A., "Schedulability analysis for real-time systems with EDF scheduling", *IEEE Transactions on Computers*, vol. 58, no. 9, pp. 1250–1258, 2009.

Index

J, K, L, M

N, O, P, Q

R, S

T, U, W

Printed in the United States
By Bookmasters